Series/Number 07-097

P9-AQI-333

ORDINAL LOG-LINEAR MODELS

MASAKO ISHII-KUNTZ
University of California, Riverside

SAGE PUBLICATIONS
International Educational and Professional Publisher
Thousand Oaks   London   New Delhi

For information address:

SAGE Publications, Inc.
2455 Teller Road
Thousand Oaks, California 91320

SAGE Publications Ltd.
6 Bonhill Street
London EC2A 4PU
United Kingdom

SAGE Publications India Pvt. Ltd.
M-32 Market
Greater Kailash I
New Delhi 110 048 India

Printed in the United States of America

Library of Congress Catalog Card No. 89-043409

Ishii-Kuntz, Masako
    Ordinal log-linear models / Masako Ishii-Kuntz
      p. cm.—(Quantitative applications in the social sciences; 97)
    Includes bibliographical references.
    ISBN 0-8039-4376-8 (pb)
    1. Log-linear models. I. Title. II. Series: Sage university
papers series. Quantitative applications in the social sciences ; no. 97.
QA278.I74 1994
300′.1519535—dc20   93-41840

94 95 96 97 10 9 8 7 6 5 4 3 2 1

Sage Production Editor: Astrid Virding

When citing a university paper, please use the proper form. Remember to cite the current Sage University Paper series title and include the paper number. One of the following formats can be adapted (depending on the style manual used):

(1) ISHII-KUNTZ, M. (1994) Ordinal Log-Linear Models. Sage University Paper series on Quantitative Applications in the Social Sciences, 07-097. Thousand Oaks, CA: Sage.

OR

(2) Ishii-Kuntz, M. (1994). *Ordinal log-linear models* (Sage University Paper series on Quantitative Applications in the Social Sciences, series no. 07-097). Thousand Oaks, CA: Sage.

CONTENTS

SERIES EDITOR'S INTRODUCTION

Suppose a contingency table analyst wishes to explore the relationship between two categorical variables, X and Y, in a 2×2 table. Traditionally, he or she might look at the percentage difference in the Y response, moving across the categories of X. For example, imagine political scientist Professor Jill Green has survey data ($N = 1,000$ voters) on college attendance (yes or no) and presidential campaign interest (yes or no). "Is education related to interest?" Apparently, for the data show that 70% of the college educated are interested, but only 40% of the non-college educated are.

As valuable as this standard percentage difference approach can be, it does not provide the answers that log-linear modeling does. Now imagine that Professor Green asks her research question a somewhat different way. "What are the *odds* that a college educated person will be interested, relative to the odds that a non-college educated person will be?" The odds for the 400 college-educated is 2.33 (i.e., 280 interested/120 not interested) compared to the odds for the 600 non-college educated of 0.67 (240 interested/360 not interested). These differing conditional odds suggest that the college educated are much more likely to be interested in the campaign. In fact, calculating the *odds ratio* (i.e., 2.33/0.67), she observes that for the college educated the odds of expressing campaign interest are 3.48 times greater than for the non-college educated. As Knoke and Burke observed in our first series monograph on the subject, this "odds ratio is the workhorse of log-linear models" (see Knoke & Burke, 1980, p. 10).

In the Knoke and Burke volume, the focus was on log-linear modeling for contingency tables with nominal variables. This volume, by Professor Ishii-Kuntz builds on that effort, extending coverage to ordinal log-linear models. (Another relevant monograph in the series is that by Demaris, 1992, *Logit Modeling: Practical Applications,* No. 86, which touches on material in both these papers). Data examples are well chosen, drawing from the National Survey of Families and Households.

The *row effects model* (i.e., the row variable is nominal, the column variable is ordinal) is illustrative of the technique's potential.

In her explication of that model, Ishii-Kuntz explores the relationship of ethnicity and gender ideology. She begins with the independence model, moves to the saturated model, and then presents a model in between these extremes—the row effects model. It is compared to the independence model and found clearly superior in terms of fit. Next, she interprets the odds ratios, for example, the odds of blacks holding a liberal gender ideology are about 2.28 times the odds of Hispanics. Professor Ishii-Kuntz goes on to explicate more complex models—column effects, uniform association, row and column effects, three-way tables. She ends with a helpful review of available computer software to estimate ordinal log-linear models.

—*Michael S. Lewis-Beck*
Series Editor

PREFACE

Statistical methods for categorical data have been extensively developed within the past three decades. Most of these discussions focus on log-linear analysis for nominal categorical data (e.g., Bishop, Fienberg, & Holland, 1975; DeMaris, 1992; Fienberg, 1980; Knoke & Burke, 1980). A monograph in the QASS series written by Knoke and Burke (1980), for instance, deals primarily with this topic. What distinguishes this monograph from the work of others listed above is its focus on log-linear models that can be used when one or more of the variables are *ordinal*.

The development of the ordinal log-linear models is reflected in several books and articles on this topic (e.g., Agresti, 1980, 1984; Goodman, 1979, 1981a, 1981b, 1981c; Hagenaars, 1990). Because the treatment of these models tends to be highly mathematical and technical, however, applied social scientists may not find them particularly helpful. Indeed, except for social and occupational mobility research (e.g., Landale & Guest, 1990; Wong, 1990; Yamaguchi, 1987), ordinal log-linear models are yet to be widely used by social scientists.

In this monograph, I have placed more emphasis on the application and interpretation of the methods than on technical details regarding estimation procedures. The monograph is divided into three sections. In Chapter 1, the concept and measurement of ordinal variables will be reviewed. A brief overview of nominal log-linear analysis is also presented. Several basic concepts such as testing goodness of fit of models, nested models, difference of chi-square tests, and odds ratios will be discussed.

Chapter 2, the main chapter of this monograph, presents various ordinal log-linear models in detail. These models include row effects, column effects, uniform association, and uniform interaction models. Ordinal log-linear models for higher ordered tables will also be described in this chapter. In addition, log-multiplicative models that do not require assignment of scores to ordinal variables are presented. To demonstrate how these models can be applied, examples using the National viii

Survey of Families and Household (NSFH) data (Sweet, Bumpass, & Call, 1988) will be presented. The NSFH was funded by a grant from the Center for Population Research of the National Institute of Child Health and Human Development. In these examples, an ordinal variable (gender ideology) will be separately examined in relation to other ordinal (religiosity and sharing of housework) and nominal (ethnicity) variables. In each example, I will discuss the fit of the model, comparison among alternative models, and odds ratios. Chapter 2 ends with a discussion of the advantages of ordinal log-linear models.

Chapter 3, the conclusion, points out additional concerns in using ordinal log-linear models, and lists some recommended references. The majority of readers will perhaps use the existing computer programs rather than write their own program for model estimation. Thus, the computer packages and their sample programs used to estimate ordinal log-linear models are given in the Appendix. Tables are numbered sequentially within chapters. Table 2.1, for example, is the first table in Chapter 2.

The final product of this monograph benefited greatly from the comments and suggestions given by a number of people. I would especially like to thank J. Scott Long, whose critical comments have significantly improved the manuscript. I am also grateful to Robert Hanneman, Susan Brumbaugh, Hisako Matsuo, Cheoleon Lee, Michael Lewis-Beck (the series editor), and two anonymous reviewers whose comments contributed to the accuracy and clarity of the monograph. I alone remain responsible for the remaining errors.

ORDINAL LOG-LINEAR MODELS

MASAKO ISHII-KUNTZ
University of California, Riverside

1. INTRODUCTION

In the past decade, social scientists have become increasingly familiar with log-linear models for *nominal* categorical variables. These models examine relationships among categorical variables by analyzing observed data. Using these models, past studies analyzed relationships among such nominal variables as religious affiliation, sex, ethnicity, and political party identification. Although these models for nominal data are now widely used, social scientists are still unfamiliar with log-linear models for ordinal variables.

With the nominal log-linear model, no assumptions are made about the order of measurement of the variables. However, the categories of some variables may have an intrinsic order. For example, political ideology may be classified as "liberal," "moderate," or "conservative"; attitudes toward abortion may be classified as "strongly approve," "approve," "disapprove," or "strongly disapprove"; and marital happiness may be classified as "very happy," "pretty happy," or "not happy." Because nominal log-linear models are insensitive to the ranking of these ordinal variables, they ignore important information when at least one variable is ordinal.

This monograph focuses on log-linear models that have been developed to analyze ordinal variables. Ordinal log-linear models are relatively simple extensions of nominal log-linear models. When nominal variables X and Y are examined in log-linear analysis, the saturated model includes the interaction term between X and Y. However, the saturated model that imposes no restrictions upon the relations between the variables has no degrees of freedom. In addition, the saturated model may be of little importance since we are primarily interested in testing more parsimonious models. The next parsimonious nominal log-linear model for a two-way table is an independence model that assumes that X and Y are unrelated. This independence model, however, is frequently unrealistic and the fit of the model is usually poor. With nominal log-linear analysis for a two-way table, we are left with no other models between independence and saturated models. The ordinal approach provides, on the other hand, an additional model between independence and saturated models that includes an X-Y association term. With ordinal models, therefore, we can test a greater variety of substantively important models. This will be illustrated in the next chapter. In this chapter, I will discuss the definition of ordinal measures and briefly review log-linear models for nominal variables.

Ordinal Measures

Ordinal measures are variables whose attributes may be logically *rank-ordered*. The different attributes represent "greater than" and "less than" rankings of the variable. One common example of an ordinal measure is social class. We can classify people into lower to middle to upper class categories. The distinction of an ordinal measure from interval and ratio measures is that although we know the ranks, we lack information on how far apart ordinal categories are. Using an ordinal arrangement, therefore, it would be irrelevant to know how close or far apart the upper and middle (or middle and lower) classes are.

Given this definition of an ordinal measure, it is important to discuss its implications for research. First, all statistical techniques require variables that meet minimum levels of measurement. Social scientists have frequently used interval level techniques to analyze ordinal variables. For example, in assessing the level of marital happiness (i.e., an ordinal variable) many family researchers have used multiple regression analysis. Using multiple regression for ordinal variables is problematic because it violates some important assumptions. For example, regression models assume homoscedasticity in the error variance of the dependent variable. This is an unrealistic assumption for ordinal categorical variables. When gross heteroscedasticity occurs (as is common with an ordinal dependent variable), the usual F-tests in regression analysis may be affected in both the true level of significance and the power of the test (see Pindyck & Rubinfeld, 1981, and Hanusheck & Jackson, 1977, for detailed discussions on problems associated with heteroscedasticity). Parameter estimates and the coefficient of determination (R^2) are also inefficient when this assumption is violated. Moreover, the use of categorical ordinal variables in regression models may yield "impossible" predicted values for the dependent variable. For example, predicted values for an ordinal variable with categories coded 1, 2, and 3 may well be outside the 1-3 range.

The frequent use of interval level analyses is partly due to the fact that social scientists have had to manage with methods designed for use in the natural sciences in which the variables are well defined and more precisely measured. In the social sciences, however, the adoption of interval level techniques has frequently resulted in a use of the analysis that does not necessarily match the level of measurement in the data. To the extent that the variables to be examined are limited to a particular level of measurement such as ordinal, it is important to plan the analytical techniques accordingly. More specifically, we should draw research conclusions appropriate to the levels of measurement used in our variables.

Second, it is important to realize that some variables may be treated as representing different levels of measurement. Ratio measures are the highest level, descending through interval and ordinal to nominal, the lowest level of measurement. A variable representing a given level of measurement—interval—may also be treated as representing a lower level of measurement—ordinal. For example, education measured by years of schooling is an interval variable. If we wish to examine only the relationship between education and some ordinal-level variable— say, marital happiness—we might choose to treat education as an ordinal variable by classifying educational attainment as "high school or less," "high school completed," "some college," and "college completed." These categories will not specify exact years of completed education and thus they are no more than a classification that reflects ordering.

Although ordinal scales can be derived from variables measured at a higher level by collapsing categories, one has to be cautious whenever this operation is attempted. In the initial data collection, the study should be designed to achieve the highest level required because the lower level of measurement can be easily constructed for later analysis. At the same time, the researcher must be aware of the consequences of collapsing categories to create lower level measures. When we collapse interval measures to the ordinal level, such as the above example of education, there is an inevitable loss of potentially useful information. For example, if the first category of educational attainment is "high school or less," we are making no distinction between individuals who completed high school and those who dropped out of high school in terms of educational achievement. In collapsing years of education, therefore, we are unable to tap some potentially interesting differences among those with and without a high school diploma.

Education is a concept that has been measured either at the ordinal or interval level. There are numerous concepts, however, that can be measured most appropriately by ordinal categories. Such concepts include satisfaction with various domains of life, level of self-esteem, and many variables measuring attitudes toward social issues. The life satisfaction variable usually consists of categories ranging from "very dissatisfied," to "very satisfied." Attitudes toward homosexuality can vary from "strongly disapprove" to "strongly approve." The log-linear models presented in this monograph are most applicable for these ordinal concepts that can be best measured by ordinal attributes.

Log-Linear Models for Nominal Variables: A Review

This section reviews the log-linear models for nominal variables. This is intended as a brief review for readers who are familiar with log-linear models. More extensive coverage and application of log-linear models for nominal variables is given by others (e.g., Agresti, 1990; Bishop, Fienberg, & Holland, 1975; Fienberg, 1980; Knoke & Burke, 1980). Although we focus on nominal log-linear models in this chapter, discussions concerning goodness of fit, nested models, and parameter estimates are directly applicable to ordinal log-linear methods. This review focuses on how log-linear models can be employed in social science research rather than providing mathematical proofs of the models. Readers who are interested in these proofs should refer to the articles or books that deal more directly with the mathematics of log-linear models (e.g., Agresti, 1990; Fienberg, 1980).

A variety of social science data come in the form of cross-classified tables of counts, commonly referred to as *contingency tables*. Until the early 1970s, the statistical and computational techniques available for the analysis of contingency tables were quite limited. Most researchers analyzed multidimensional contingency tables by studying two variables at a time and using available computer program packages that produced chi-square statistics. Although such an approach often gives great insight about the relationship among variables, it has its limitations. For example, the traditional approach does not allow for the simultaneous examination of pairwise relationships, and it ignores the possibility of three-factor and higher-order interactions among the variables. Log-linear models take care of the shortcomings mentioned above. With the log-linear approach, we model cell counts in a contingency table in terms of associations among the variables and marginal frequencies. In this review of nominal log-linear models, we analyze the relationship between sex and whether or not a respondent received emotional support from friends during the month prior to the survey. Intuitively, we would predict that women receive more emotional support from their friends than their male counterparts because our society stereotypes women to be more dependent and sociable. There is also little doubt that men and women have different socialization experiences throughout their lives, with females receiving greater reinforcement during socialization for expressive skills. Boys' peer groups, on the other hand, often focus on instrumental tasks rather than socioexpressive ones. It is not surprising, therefore, that Keith, Hill, Goudy, and Powers (1984) found that men have lower expectations of intimacy and self-disclosure than do women. Based on this theoretical premise and research findings, we hypothesize that women are more likely to receive emotional support from friends than are men.

To test this hypothesis, we use data from the National Survey of Families and Households (NSFH) conducted in 1987-1988 by the Center for Demography and Ecology, the University of Wisconsin, Madison (Sweet, Bumpass, & Call, 1988). The main sample contains about 17,000 housing units drawn from 100 sampling areas in the coterminous United States. One adult aged 19 and older in each household was randomly chosen for an interview and questionnaire survey. The subsample used for the present analysis contains married respondents ($N = 6,373$).

Table 1.1 is a contingency table that shows the observed frequencies (f_{ij}'s) at the various combinations of sex and emotional support variables. Table 1.1 has two dimensions, i (1,2) rows that represent the categories of sex and j (1,2) columns that represent the categories of a second variable, emotional support.

Log-Linear Model for Two-Way Tables

We first test a model that posits that one's gender has no relationship with whether or not one receives emotional support from friends. Table 1.2 shows the expected cell frequencies (F_{ij}) under the model of inde-

TABLE 1.1

Cross-Tabulation of Sex [S] and Emotional Support [E]

Sex	Emotional Support		
	Yes	*No*	*Total*
Male	785	2,119	2,904
Female	1,416	2,053	3,469

pendence for our example. These expected frequencies mean that if the hypothesized independence model were true, we would have obtained these frequencies instead of the observed frequencies in Table 1.1.

A log-linear model can be developed from these expected frequencies. Taking the natural logarithm of the equation for expected frequencies, $F_{ij} = (f_{i\cdot} f_{\cdot j})/N$, an additive relationship in the logs can be expressed as follows:

$$\log F_{ij} = \log f_{i\cdot} + \log f_{\cdot j} - \log N \tag{1.1}$$

where dot notations denote summation over the rows or columns. This equation shows that if two variables are independent, the log of the expected frequency for cell (i, j) is an additive function of an ith row "effect" and a jth column "effect." An alternative formulation of Model 1.1 for sex (S) and emotional support (E) example is:

$$\log F_{ij} = \mu + \lambda_i^S + \lambda_j^E . \tag{1.2}$$

This model is called the *log-linear model for independence* in a two-way table, where μ is the grand mean of the logarithms of the expected counts, λ_i^S is the "effect" of sex, and λ_j^E is the "effect" of emotional support. Within log-linear models, labeling the λ parameters "effects" or "effect parameters" does not have any causal connotation. All that is implied by the term *effect* is that the sizes of the cell frequencies are a mathematical function of the sizes of the λ parameters. Using Goodman's (1979) shorthand notation of fitted marginals, this model can also be expressed as [S][E]. Because the independence model posits gender and emotional support to be unrelated to each other, the initials of each variable appear in separate braces.

Expected Frequencies Under the Model of Independence
for Sex and Emotional Support

Sex	Emotional Support		Total
	Yes	*No*	
Male	$F_{11} = \dfrac{(f_{1.})(f_{.1})}{N}$	$F_{12} = \dfrac{(f_{1.})(f_{.2})}{N}$	2,904
	$= 1,002.9$	$= 1,901.1$	
Female	$F_{21} = \dfrac{(f_{2.})(f_{.1})}{N}$	$F_{22} = \dfrac{(f_{2.})(f_{.2})}{N}$	3,469
	$= 1,198.1$	$= 2,270.9$	

The log-linear independence model (1.2) includes no constraints. Without additional constraints, the log-linear model is overparameterized—there are more parameters than cells in the table. In an overparameterized model individual parameters cannot be estimated (see Long, 1984, for more discussion on estimable parameters). Therefore, for the purpose of estimation, constraints must be imposed. Paralleling the ANOVA model, the constraints that the sum of the effects of sex is zero and that the sum of the effects of emotional support is zero $(\sum_i \lambda_i^S = \sum_j \lambda_j^E = 0)$ are frequently imposed. These will be called ANOVA-like (or zero-sum) constraints. Due to these constraints, there are $I - 1$ linearly independent row effect parameters and $J - 1$ linearly independent column effect parameters where I and J are number of row and column categories, respectively. Thus the independence model described in (1.2) has $1 + (I - 1) + (J - 1) = I + J - 1$ linearly independent parameters. To compute the degrees of freedom associated with log-linear models, the number of cells in the contingency table must be known. A saturated model always has no available degrees of freedom because all conceivable parameters are free to vary in fitting the data precisely. As the number of parameters to be estimated from the data are reduced, degrees for testing the model are increased by the equivalent number.

For an independence model with sex and emotional support variables, we obtain: $\hat{\mu} = 7.319$, $\hat{\lambda}_1^S = -0.089$, $\hat{\lambda}_2^S = 0.089$, $\hat{\lambda}_1^E = -0.320$, and

$\hat{\lambda}_2^E = 0.320$. Substantively, these estimates are no more than the deviation from the overall mean. For example, -0.089 ($\hat{\lambda}_1^S$) indicates that this expected frequency is slightly smaller than that of the entire table. On the contrary, the average expected frequency for females (0.089) is slightly larger than that of the overall table. However, these interpretations are not of substantive importance for many researchers.

Odds and Odds Ratios

How do we interpret log-linear parameters in a way that makes more sense? Log-linear modeling requires a very different way of viewing relationships among variables from what many social scientists are accustomed to. The "effects" or the degrees of association of the log-linear models are interpreted in terms of odds and odds ratios. Odds are defined as the frequency (or probability) of one category of a variable compared to the frequency (or probability) of another. Odds are unbounded at the upper end; that is, they can take on any value greater than 0. Suppose we want to know the odds of a respondent being a male as opposed to a female; we simply divide the number of males by the number of females. In this case, the value of odds that are greater than 1.0 means that respondents are more likely to be males than females, and the odds that are less than 1.0 means that respondents are more likely to be females than males. If the odds of the respondent being male as opposed to female are 2, then it means that the respondent is twice as likely to be a male as a female. On the other hand, if the same odds are 0.5, it means that the respondent is twice as likely (the inverse of 0.5) to be a female as a male.

Interpretation of the log-linear parameters using odds is often easier when we use an exponential transformation of the log-linear model. Taking the exponent of (1.2) results in a *multiplicative model:*

$$e^{\log F_{ij}} = e^{(\mu + \lambda_i^S + \lambda_j^E)} \tag{1.3}$$

$$= e^\mu \, e^{\lambda_i^S} \, e^{\lambda_j^E}$$

Expressed differently, (1.3) becomes:

$$e^{\log F_{ij}} = \eta \, \pi_i^S \, \pi_j^E \tag{1.4}$$

where $\eta = e^\mu$, $\pi_i^S = e^{\lambda_i^S}$ and $\pi_j^E = e^{\lambda_j^E}$. Model 1.4 is a model that is multiplicative in the expected frequencies rather than linear in the log of the expected frequencies. The exponential transformation also affects the functions that are estimated in the multiplicative model. For example, if in the log-linear formulation of a model the function $[\lambda_1^S - \lambda_2^S]$ is estimated, in the multiplicative formulation the function $e^{[\lambda_1^S - \lambda_2^S]} = \pi_1^S/\pi_2^S$ is estimated. With the multiplicative model, parameter estimates have more straightforward interpretations because they are presented in an "odds" form.

The exponential transformation of the parameter estimates in our example yields: $\hat\eta = 1508.695$, $\hat\pi_1^S = 0.915$, $\hat\pi_2^S = 1.093$, $\hat\pi_1^E = 0.726$, and $\hat\pi_2^E = 1.377$. Using these multiplicative parameters, the odds for being a female as opposed to a male can be obtained by the function $[\hat\pi_2^S/\hat\pi_1^S] = 1.093/0.915 = 1.195$. This means that *if sex is independent of emotional support from friends,* the odds of respondents being female as opposed to male is slightly greater. Similarly, the odds of not receiving (versus receiving) emotional support from friends is $[\hat\pi_2^E/\hat\pi_1^E] = 1.377/0.726 = 1.897$. Therefore, the respondents are almost twice as likely to report that they received no emotional support from friends. These interpretations of odds make more intuitive sense than the "difference" interpretation of two log-linear parameter estimates.

A log-linear model is called hierarchical when any parameter is set to 0, all effects of the same and higher order are set to zero. For example, if in a model for a four-dimensional table $A \times B \times C \times D$, the parameter λ_{ij}^{AB} is set to zero, the hierarchy principle implies that λ_{ijk}^{ABC}, λ_{ijl}^{ABD}, and λ_{ijkl}^{ABCD}, each containing the superscript AB, are also set to zero. For *nominal* hierarchical log-linear models of two-way tables, the model of the next higher complexity is the *saturated model.* In a saturated model no restrictions are imposed upon the relations between the variables, whereas in an unsaturated model (such as the independence model), the relations between the variables are restrained in some way, mostly by assuming that one or more effects between variables are absent, that is, λ_{ij}^{SE} in the independence model. For a two-way table in our example, the saturated model is expressed:

$$\log F_{ij} = \mu + \lambda_i^S + \lambda_j^E + \lambda_{ij}^{SE} \qquad (1.5)$$

or [S][E][SE]. In Model 1.5 the additional λ_{ij}^{SE} parameters are association parameters that reflect departures from the independence model. That is, inclusion of the λ_{ij}^{SE} term indicates that sex and emotional support are now assumed to be related. These parameters satisfy zero-sum constraints; $\sum_i \lambda_i^S = \sum_j \lambda_j^E = 0$ and $\sum_i \lambda_{ij}^{SE} = \sum_j \lambda_{ij}^{SE} = 0$. The multiplicative form of this model can be expressed:

$$F_{ij} = \eta \pi_i^S \pi_j^E \pi_{ij}^{SE}$$

Although this model is saturated, a substantively more interesting interpretation of the parameters can be generated using odds ratios: What are the odds of females receiving emotional support from friends compared to the odds of males?

To answer this question, we first compute individual parameter estimates (λ's) and exponentially transpose them into π's: $\hat{\pi}_1^S = 0.870$, $\hat{\pi}_2^S = 1.150$, $\hat{\pi}_1^E = 0.711$, $\hat{\pi}_2^E = 1.407$, $\hat{\pi}_{11}^{SE} = 0.856$, $\hat{\pi}_{12}^{SE} = 1.168$, $\hat{\pi}_{21}^{SE} = 1.168$, and $\hat{\pi}_{22}^{SE} = 0.856$. Then we compute odds and odds ratio as follows:

STEP 1: What are the odds of females receiving emotional support from friends?

$$(\hat{\pi}_1^E / \hat{\pi}_2^E)(\hat{\pi}_{21}^{SE} / \hat{\pi}_{22}^{SE}) = (0.711/1.407)(1.168/0.856)$$

$$= (0.5053)(1.3644) = 0.6894$$

STEP 2: What are the odds of males receiving emotional support from friends?

$$(\hat{\pi}_1^E / \hat{\pi}_2^E)(\hat{\pi}_{11}^{SE} / \hat{\pi}_{12}^{SE}) = (0.711/1.407)(0.856/1.168)$$

$$= (0.5053)(0.7329) = 0.3703$$

STEP 3: What are the relative odds (odds ratio) of females receiving emotional support from friends versus the males receiving emotional support from friends?

$$\frac{(\hat{\pi}_1^E / \hat{\pi}_2^E)(\hat{\pi}_{21}^{SE} / \hat{\pi}_{22}^{SE})}{(\hat{\pi}_1^E / \hat{\pi}_2^E)(\hat{\pi}_{11}^{SE} / \hat{\pi}_{12}^{SE})} = \frac{1.3644}{0.7329} = 1.862$$

The odds ratio of 1.862 means that the odds of females receiving support is about 1.9 times the odds of males receiving support, thus our hypothesis is supported.

It is also important to know that upon simplification, the above odds ratio becomes the cross-product ratio for a 2×2 table:

$$\text{Cross Product Odds Ratio} = \Theta = \frac{\pi_{21}^{SE} \pi_{12}^{SE}}{\pi_{11}^{SE} \pi_{22}^{SE}}$$

For the saturated model, we can also obtain this cross-product ratio from the observed frequencies, $[f_{21} f_{12} / f_{11} f_{22}] = [1,416 \times 2,119] / [785 \times 2,053] = 1.862$.

In summary, we use odds and odds ratios to interpret log-linear parameters and the odds are easily computed using the multiplicative parameters. One confusing aspect of log-linear modeling, however, is that some widely used computer programs (such as BMDP4F, SAS-CATMOD, and SPSS-LOGLINEAR) do not provide the parameters that are readily interpretable for odds and odds ratios. BMDP4F, for example, gives estimates of multiplicative parameters. It does not, however, compute the ratio of the multiplicative parameter estimates. Interpreting odds ratios in log-linear models may thus require some computation using the numbers in the computer output. For more discussion on interpretation of log-linear parameters, refer to Alba (1988), Clogg and Eliason (1988), and Kaufman and Schervish (1986).

Assessing the Fit of Log-Linear Models

We have so far discussed the interpretation of parameter estimates generated by log-linear models assuming that the hypothesized model fits the observed data well. But how do we assess the fit of the model? Let us now discuss how to evaluate whether a hypothesized log-linear model fits the observed data. This process involves estimating the expected cell frequencies, F_{ij}'s, for the proposed model and comparing them to the observed frequencies, f_{ij}'s, using either the *Pearson chi-square statistic* (χ^2) or the *likelihood-ratio statistic* (G^2). In general, G^2 is preferable to χ^2 because the G^2 is statistically minimized under maximum likelihood methods, and G^2 can be conditionally broken down to allow likelihood ratio tests of nested models. However, when the independence model is true, both χ^2 and G^2 have asymptotic (as $n \to \infty$) chi-squared distributions with degrees of freedom df $= (I-1)(J-1)$. That is, χ^2 and G^2 are equivalent in very large samples when H_0 is true.

Various guidelines have been provided for how large the sample size should be in order for the chi-squared distribution to have a good approximation for the exact sampling distributions of the χ^2 and G^2 statistics. A commonly quoted guideline follows Cochran (1954), who suggests that at least 80% of the cells should have F_{ij} exceeding 5.0, and F_{ij} should exceed 1.0 in all cells. Larntz (1978) and Koehler and Larntz (1980), however, showed that the chi-squared approximation can be very good for the χ^2 statistic even for very small expected frequencies.

For either chi-square statistic (χ^2 or G^2) larger values provide more evidence against the null hypothesis. The attained significance level is the probability of getting a statistic value larger than the observed one, assuming H_0 is true. One note of caution in interpreting chi-squared statistics is that we are always testing against the null hypothesis. Thus if the model posits independence between two variables, then the larger chi-square and smaller probability is a sign that we reject the null hypothesis of independence. For the independence model of gender and emotional support, we obtain $G^2 = 134.35$ with one degree of freedom. This means that we reject the hypothesized independence model and conclude that whether or not one receives emotional support from friends varies depending on gender.

For more complex models, we wish to accept the null hypothesis that posits the proposed model. For example, in a three-dimensional contingency table, $A \times B \times C$, if the hypothesized model has all second-order terms [AB][AC][BC], then the smaller chi-square with a larger probability indicates that we *fail* to reject the null model with all three interaction terms. Subsequently, we accept the proposed model that posits that each variable is related to another variable but there is no third-order interaction term [ABC].

Partitioning G^2 for Nested Models

As in regression analysis, a variety of models can be applied to contingency tables. And, as in regression analysis, these models can be tested against one another. Comparing models can be done statistically if the models are nested. Consider independence (1.2) and saturated (1.5) models in our sex and emotional support example:

$$\log F_{ij} = \mu + \lambda_i^S + \lambda_j^E$$

and

$$\log F_{ij} = \mu + \lambda_i^S + \lambda_j^E + \lambda_{ij}^{SE}.$$

Here the independence model is nested in the saturated model because these two models satisfy the following conditions: (a) If the independence model is the true model, the independence model implies the saturated model, but the saturated model is not as parsimonious as the independence model; (b) The independence model is a special case of the saturated model when $\lambda_{ij}^{SE} = 0$; (c) The independence model has more degrees of freedom than the saturated model; (d) G^2 for the independence model is greater than G^2 for the saturated model, which is always zero; and (e) The marginals fixed in the independence model are a subset of the marginals fixed by the saturated model.

Because G^2 can be conditionally broken down, nested models can be compared using difference of G^2 values and degrees of freedom. For our 2×2 table of sex and emotional support, we obtain $G^2 = 134.35$ with df = 1 for the independence model and $G^2 = 0$ with df = 0 for the saturated model. A saturated model in log-linear analysis has as many parameters as there are cells in the table. The saturated model hypothesizes, in effect, that the model is identical to the relationship in the observed table. Consequently, a chi-square and the degrees of freedom of the saturated model are always zero. By taking the difference of G^2's between independence and saturated models, we obtain $\Delta G^2 = 134.35$ and $\Delta df = 1$. The null hypothesis being tested by a difference of G^2's is H_0: $\lambda_{ij}^{SE} = 0$, thus with the above difference in G^2's and df, we reject the null hypothesis. This means that adding the λ_{ij}^{SE} term significantly improves the fit of the model. Stated another way, the association between sex and emotional support from friends is statistically significant. Although the example given here is a rather simple one, the difference of G^2's is a useful statistical test that can be easily applied to higher ordered tables.

Log-Line-ar Models for
Multiway Tables

Most realistic research questions involve more than two variables. We may ask, for instance, "Do males and females differ in the relationship between social support and psychological well-being?" Or stated another way, "Is the effect of social support on well-being greater for females than for males?" Further, we may want to compare these relationships over different ethnic groups. Two-way analyses are clearly inadequate to address such questions. Log-linear models allow the easy generalization of the statistical techniques for two-way tables.

Consider a three-way table (Table 1.3) with *s*ex (S), *e*motional support from friends (E), and *p*sychological well-being (P), where sex and emotional support have two categories each as specified in the previous example, and psychological well-being has three categories (low, medium, high). The saturated model for this three-way table is:

$$\log F_{ijk} = \mu + \lambda_i^S + \lambda_j^E + \lambda_k^P + \lambda_{ij}^{SE} + \lambda_{ik}^{SP} + \lambda_{jk}^{EP} + \lambda_{ijk}^{SEP}.$$

This model can be also designated as [S][E][P][SE][SP][EP][SEP] or simply [SEP] assuming that all the lower terms are included in the hierarchical log-linear model. As with the log-linear model for two-way tables, the zero-sum constraints are imposed for this model. In multiway tables such as Table 1.3, substantive interest generally focuses on special cases of the saturated model. That is, on models that are nested in the saturated model. Several examples of these hierarchical models for Table 1.3 include:

1. [SE][SP][EP]. This model fits all two-factor associations but no three-factor association with df = $(I - 1)(J - 1)(K - 1)$. In this model, all pairs of variables are related and the nature of that relationship is not dependent upon the category of the third variable. That is, sex is related to emotional support, and it is also related to psychological well-being. However, each of these bivariate relationships does not depend on the third variable. Emotional support is also related to psychological well-being regardless of one's sex.

2. [SE][SP]. This model fits two of the three associations and not the three-factor association with df = $(I)(J - 1)(K - 1)$. Sex is related to emotional support and psychological well-being separately, but no association between emotional support and psychological well-being is assumed. The variations of this model are [SE][EP] and [SP][EP].

3. [SE][P]. Sex and emotional support are related, but psychological well-being is related neither to sex nor to emotional support from friends. The degrees of freedom for this model are $(IJ - 1)(K - 1)$. The variations of this model are [SP][E] and [EP][S].

The extension of hierarchical log-linear models for higher-order tables (four-way, five-way, six-way, etc.) is straightforward, even though practical problems abound. One of the problems is that the inclusion of more variables reduces cell frequencies dramatically. For instance, a 3 × 3 table has 9 cells; a 3 × 3 × 3 table has 27 cells; a 3 × 3 × 3 × 3 table has 81 cells; and a 3 × 3 × 3 × 3 × 3 table has 243 cells. Thus, with a sample of 1,000, an average cell in the last table would have less than five observations; in practice many zero cells would be expected.

Another problem is that with the additional variables, the models quickly become very complex and thus the interpretation of the parameters may become unmanageable. In a four-way table, for instance, there

TABLE 1.3

Cross-Tabulation of Sex [S], Emotional Support [E], and
Psychological Well-Being [P]

		Psychological Well-Being		
Sex	Emotional Support	Low	Medium	High
Male	Yes	56	245	478
Male	No	139	462	1,473
Female	Yes	153	599	660
Female	No	247	667	1,101

can be one zero-order effect; four first-order effects; six second-order effects; four third-order effects; and one fourth-order effect. This allows 27 kinds of hypotheses, plus the saturated model. In a five-way table there are 32 types of effects. In a six-way table, there are 64 types of effects. In short, the objective of log-linear modeling is to come up with more parsimonious models that have a reasonable fit to the observed data rather than to come up with models with many variables that increase the problems of zero cells and make the interpretation of parameters more complicated.

Selection of Models

In contingency tables with more variables, there are hundreds of models that can be considered. Dozens may provide acceptable levels of fit at conventional levels of significance. How then does one find "the best fitting" model? In an ideal world one's theory suggests a model, it is tested, and the appropriate conclusion is drawn. More realistically, substantive theory usually provides only partial guidance about which model should fit the data. And, substantive theory may suggest a specific model, but that model might not fit. Most users of log-linear models find themselves in a specification search, a formalized strategy for searching through possible models. There are three primary processes in all specification searches.

1. Selecting a Baseline Model. The baseline model is the first model that is examined. The baseline model could be a model suggested by theory, the saturated model, the zero-order model, or a model of order *k*.

2. Modifying the Baseline Model. After the baseline model is chosen, some systematic set of procedures is needed. This involves deciding which terms to add to and/or delete from the baseline model. This process involves forward or backward selection of parameters.

3. Stopping the Specification Search. This involves deciding when the model fits "well enough," or when a "best fitting" model has been found.

For the three-way relationships in Table 1.3, besides the saturated model there are 17 possible models as shown in Table 1.4. The only model that fits the data is Model 17. The comparison between Model 17 and the saturated model tests if the three-factor interaction term is significant. This difference is $\Delta G^2 = 1.11$ for $\Delta df = 2$, thus we must accept H_0: $\lambda_{ijk}^{SEP} = 0$ and conclude that although sex is related to emotional support and psychological well-being separately, and emotional support is related to well-being, there is no three-way relationship among these variables. We can also compare if the second-order terms included in Model 17 are significant. For example, the difference between Models 16 and 17 is $\Delta G^2 = 163.73$ for $\Delta df = 2$ confirming that the relationship between gender and psychological well-being is significant and thus should be included in the final model. Similar comparisons show that both [SE] and [EP] terms significantly improve the fit of the model and thus should be retained in the final model. Therefore, we choose [SE][SP][EP] model as the final model.

The selection of this model satisfies all of the conditions for the "best fitting" model. The model is parsimonious with no third-order terms, and it contains all significant terms. It is important to note, however, that these conditions are not necessarily consistent, exclusive, or exhaustive. Indeed, each researcher will tend to choose the model that "seems right." It is always important to remember that although statistics can guide the specification search, it is more art than science, even if it is common to all sciences. Thus the importance of theoretical guidance in the selection of log-linear models cannot be overemphasized.

Finally, in a specification search the results of statistical tests on a data set are used to suggest additional statistical tests on the same data. As a consequence, one should interpret the fit of the final model cautiously. To illustrate, suppose the researcher decides to begin with a model that contains all second-order terms [SE][SP][EP]. Difference of G^2 tests is used to assess each second-order term, that is, comparisons between Models 14 and 17, 15 and 17, and 16 and 17. The hypotheses

TABLE 1.4

Nominal Log-Linear Models for Data in Table 1.3

Model		G^2	df	prob
1	[S]	3,485.71	10	.0000
2	[E]	2,955.55	10	.0000
3	[P]	1,016.26	9	.0000
4	[S][E]	2,903.01	9	.0000
5	[S][P]	963.72	8	.0000
6	[E][P]	433.56	8	.0000
7	[S][E][P]	381.02	7	.0000
8	[SE]	2,769.18	8	.0000
9	[SP]	773.86	6	.0000
10	[EP]	351.20	6	.0000
11	[S][EP]	298.67	5	.0000
12	[E][SP]	191.16	5	.0000
13	[P][SE]	247.19	6	.0000
14	[SE][SP]	57.33	4	.0000
15	[SE][EP]	164.84	4	.0000
16	[SP][EP]	108.81	3	.0000
17	[SE][SP][EP]	1.11	2	.5741

tested here are H_0: $\lambda_{ij}^{SE} = 0$, or $\lambda_{ik}^{SP} = 0$, or $\lambda_{jk}^{EP} = 0$. Upon these tests, the researcher decides to keep all second-order terms that are significant. However, if a Type I error is committed (reject a true H_0) for each test, a term that should be dropped will be kept. Therefore, the final model may include irrelevant terms. In short, the specification search may result in a final model that includes erroneous terms.

Zero Cells in Contingency Tables

Although the log-linear models are versatile statistical models, there are some limitations in using them. These limitations are largely due to zero cells that may arise in contingency tables for two main reasons. *Fixed zeros* occur when it is impossible to observe values for certain combinations of the variables. For example, when we cross-tabulate sex and number pregnancies, zero-cell entry occurs for males and pregnancies. *Sampling zeros* are due to sampling variation and the relatively small size of the sample when compared with the large number of cells. The consequences of the zero-cell problem are twofold. First, we cannot include many variables in the analysis. This is also related to the second consequence: eventual collapsing of the variables into dichotomies or trichotomies in order to avoid zero cells in the table. The dilemma here is that collapsing categories or dimensions may distort the process being modeled and may result in a loss of some valuable data; but not collapsing may leave zero cells that may violate assumptions of the log-linear model. In addition, the appearance of zeros in one or more cells can be a problem, because odds and odds ratios are undefined with zeros in the denominator. In collapsing categories, researchers must be aware of these consequences of the collapsing (for more discussion, see Agresti, 1990).

Other than collapsing variable categories, several options are available when zero cells are present in contingency tables. They are: (a) Add a small value (0.5 is frequently suggested) to every cell in the table when fitting the saturated model (Goodman, 1970); (b) Add a small quantity such as 0.2 only to zero cells (Evers & Namboodiri, 1977); (c) Add the value $1/r$ to zero cells, where r equals the number of response categories (Grizzle, Starmer, & Koch, 1969); (d) Arbitrarily define zero divided by zero to be zero (Fienberg, 1980); (e) Increase the sample size sufficiently to remove all zero cells (Knoke & Burke, 1980); and (f) Replace sampling zeroes by 0.1×10^{-8}, or a smaller number and then check results against those obtained without such an adjustment (Clogg & Eliason, 1988).

In summary, researchers need to consider carefully the limitations of the log-linear models in order to analyze the categorical data effectively. Up to now, we have only described the log-linear models for nominal variables. In the following chapter, we present log-linear models that are specifically developed to analyze ordinal variables.

2. ORDINAL LOG-LINEAR MODELS

All the models that have been reviewed to this point make no assumptions about the order of the variable categories. If one or more variables are ordinally measured, we can use a variety of log-linear models that take into account the ordinal nature of the variables. This chapter discusses these ordinal log-linear models.

We begin by discussing ordinal log-linear models for two-way tables that treat one variable as ordinal and the other as nominal. Two variations of these models are discussed here: the *row effects model* and the *column effects model.* In row effects models, the nominal variable is the row variable and the ordinal variable is the column variable. In the column effects model, the ordinal variable is the row variable and the nominal variable is the column variable. Second, a model that treats both variables as ordinal is discussed. Finally, ordinal log-linear models for higher ordered tables are presented.

Row Effects Models

To illustrate how row effects models can be used, let us consider a relationship between ethnicity [E] and attitudes toward men's and women's roles (we will call it *gender ideology*) [G] where ethnicity is a nominal variable and gender ideology is an ordinal variable. The gender ideology is measured by the extent of agreement/disagreement with the statement, "It is much better for everyone if the man earns the main living and the woman takes care of the home and family." The categories of the gender ideology range from "traditional" (those who agreed with the above statement) to "moderate" (those who checked midpoint between agreement and disagreement) to "liberal" (those who disagreed with the above statement). Table 2.1 shows the relationship between these two variables for a sample of *married* respondents in the 1988 National Survey of Families and Household data (NSFH).

The independence model for the two-way cross-classification in Table 2.1 is:

$$\log F_{ij} = \mu + \lambda_i^E + \lambda_j^G.$$

We do not expect the independence model of ethnicity and gender ideology to give an adequate fit of the data. This prediction is based on prior research that found that the division of family responsibilities into stereotypically male versus female roles is substantially less characteristic of black families than white families (Beckett & Smith, 1981). Hispanic families, on the other hand, tend to be more traditional than their white counterparts with respect to gender ideology (Williams, 1990; Zavella, 1987). It is, therefore, predicted that blacks report a more liberal gender ideology, Hispanics hold more traditional gender ideology, and whites' attitudes toward women's roles tend to be somewhere between traditional and liberal. In other words, the association term between ethnicity and gender ideology needs to be incorporated in the

TABLE 2.1

Observed and Expected Frequencies for Independence and Row
Effects Models of Ethnicity [E] and Gender Ideology [G]

Ethnicity	Gender Ideology			
	Traditional	Moderate	Liberal	Total
Black	73	67	77	217
	(76.2)[a]	(74.9)	(65.9)	
	[68.5][b]	[75.9]	[72.5]	
White	464	510	443	1,417
	(497.4)	(489.5)	(430.1)	
	[470.9]	[496.1]	[449.9]	
Hispanic	106	47	23	176
	(62.5)	(60.7)	(52.8)	
	[104.1]	[50.7]	[21.1]	
Total	643	624	543	1,810

NOTES: a. For independence model F_{ij}.
    b. For row effects model F_{ij}.

proposed model to test if the relationship between two variables is
significant.

In the hierarchical nominal log-linear modeling approach, the model
that includes the association term is saturated:

$$\log F_{ij} = \mu + \lambda_i^E + \lambda_j^G + \lambda_{ij}^{EG} .$$

In this saturated model, no assumptions are made about the causal
structure of the data. The expected frequencies from a saturated model
always perfectly match the observed frequencies. Because the saturated
model only reproduces the observed association in the table and thus
has zero degrees of freedom, there is no way of telling "how good" the
saturated model is in terms of goodness of fit.

By introducing notions of ordering, we can find a model that falls
between independence and saturated models. That is, if one of the
variables is ordinal, a *nonsaturated* log-linear model that includes the
two-factor association term, [EG], can be used. This model is known as
row effects model and it is constructed by replacing λ_{ij}^{EG} in the saturated
model with an association term $\tau_i (v_j - \bar{v})$:

$$\log F_{ij} = \mu + \lambda_i^E + \lambda_j^G + \tau_i (v_j - \bar{v}) \qquad (2.1)$$

where the τ_i's are row effects parameters and the v_j are assigned scores to the columns in the contingency table where $v_1 < v_2 < \ldots < v_J$. In many applications the choices of scores will be based on assumed distances between midpoints of categories for the underlying interval scale. Equally spaced scores result in the simplest interpretation for the model discussed here. In addition, the usual zero-sum constraints $\sum \lambda_i^E = \sum \lambda_j^G = \sum \tau_i = 0$ are imposed in order to identify the model's parameters.

Including fixed constants, $(v_j - \bar{v})$, in the association term is a restriction imposed on the two-factor association term in the saturated model. Because of this restriction, there are now degrees of freedom to test the fit of the model. That is, Model 2.1 has $I - 1$ more parameters (the τ_i) than the independence model, thus its degrees of freedom are $IJ - [1 + (I - 1) + (J - 1) + (I - 1)] = (I - 1)(J - 2)$. Therefore, the row effects model is more parsimonious than the saturated model.

In the row effects model, μ and the λ parameters are the usual log-linear terms for the overall mean and the main "effects" of ethnicity and gender ideology. What differs is the inclusion of the term involving row effects parameters τ_i. The fixed constants, $(v_j - \bar{v})$, are equal to scores $(-1, 0, +1)$ that correspond to the three categories of an ordinal column variable, reflecting the rank-ordered categories of gender ideology. It should be clear that the association term, $\tau_i (v_j - \bar{v})$, reflects a deviation from the independence model, and the deviation increases in the direction of scores of the ordinal column variable alone. That is, within a given row, a positive τ indicates that more observations occur in columns representing high values of the ordinal variable, and fewer in columns representing low values of the variable, compared to what would be expected under the independence model. In our example, τ's indicate whether blacks (or Hispanics or whites) are more or less likely to have a traditional or liberal gender ideology. Using this example, we will first describe the fit of the model.

Assessing Goodness of Fit

In Table 2.1 observed and expected frequencies for the independence and the row effects models are presented. Marginal frequencies show that more people felt that women should stay home and avoid paid employment. Although black and white respondents seem to be split

TABLE 2.2
Goodness-of-Fit Models for Ethnicity and Gender Ideology Data

Model	G^2	df	prob
Independence	57.64	4	.000
Row Effects	2.70	2	.259

across categories of gender ideology, Hispanic respondents tend to hold more traditional attitudes. We compare the independence model that posits no relationship between ethnicity and gender ideology with the row effects model that hypothesizes that gender ideology varies depending upon a person's ethnic backgrounds.

The goodness of fit of the independence model and of the row effects model is summarized in Table 2.2. As predicted, the fit of the independence model is poor with $G^2 = 57.64$ based on df = 4. With the addition of the row effect parameters (τ_i) the fit of the model is significantly better $(G^2 = 2.70, df = 2)$, indicating a strong association between ethnicity and gender ideology. The difference of the chi-square test between independence and row effects models results in $\Delta G^2 = 54.94$ with $\Delta df = 2$. We thus conclude that including row effects parameters significantly improves the fit of the model. In other words, there is a significant difference in gender ideology among blacks, whites, and Hispanics.

Odds Ratios

After examining the fit of the model, the next step is to interpret the parameter estimates of the row effects model. That is, given the significant association between ethnicity and gender ideology, we ask *how* gender ideology varies among the three groups or *how strong* the association is between ethnicity and gender ideology. We use odds ratios to describe row effects parameters. The odds ratios are formed using cells in adjacent rows and adjacent columns. For an arbitrary pair of rows h and i and adjacent columns j and $j + 1$, the log odds ratio (log Θ_{ij}) is:

$$\log \frac{F_{hj} F_{i,j+1}}{F_{h,j+1} F_{ij}} = \log (F_{hj} F_{i,j+1}) - \log (F_{h,j+1} F_{ij})$$

$$= [\log (F_{hj}) + \log (F_{i,j+1})] - [\log (F_{h,j+1}) + \log (F_{ij})]$$

$$= [\mu + \lambda_h + \lambda_j + \tau_h (v_j - \bar{v}) + \mu + \lambda_i + \lambda_{j+1} + \tau_i (v_{j+1} - \bar{v})]$$

$$- [\mu + \lambda_h + \lambda_{j+1} + \tau_h (v_{j+1} - \bar{v}) + \mu + \lambda_i + \lambda_j + \tau_i (v_j - \bar{v})]$$

$$= [\tau_h (v_j - \bar{v}) + \tau_i (v_{j+1} - \bar{v})] - [\tau_h (v_{j+1} - \bar{v}) + \tau_i (v_j - \bar{v})]$$

$$= \tau_h (v_j - v_{j+1}) + \tau_i (v_{j+1} - v_j)$$

$$= \tau_i - \tau_h \tag{2.2}$$

because $(v_j - v_{j+1}) = -1$ and $(v_{j+1} - v_j) = 1$. We use an exponential transformation of these log odds ratios to obtain odds ratios, thus:

$$\Theta_{ij} = e^{\tau_i - \tau_h}.$$

In the row effects model, this odds ratio is identical for all combinations of adjacent columns. That is, for a given pair of rows (h and i) the odds ratios for any adjacent columns (j vs. $j + 1$), are identical. The ($\tau_i - \tau_h$) describes differences among rows with respect to their conditional distributions on the column variable. When τ_i and τ_h are equal, rows i and h have identical conditional distributions on the column variable. When τ_i is greater than τ_h, then the subjects in row i are more likely to respond to higher categories on the ordinal scale compared to the subjects in row h.

In the ethnicity and gender ideology example, the estimated $\hat{\tau}$'s are $\hat{\tau}_1 = .292$ for black, $\hat{\tau}_2 = .241$ for white, and $\hat{\tau}_3 = -.533$ for Hispanic. The more positive τ is, the greater the tendency for respondents to be at the more liberal end of the gender ideology scale. The $\hat{\tau}_1$ and $\hat{\tau}_2$ are positive, indicating that blacks and whites are more likely to hold a liberal gender ideology than the independence model would predict. The $\hat{\tau}_3$ is negative, indicating that there is an excess of Hispanic respondents who classify themselves into the traditional category of gender ideology.

We then estimate *how much* more likely are blacks (or whites or Hispanics) to have a liberal (or moderate or traditional) gender ideology compared to respondents in other race categories. As shown above, the comparison between groups in terms of their gender ideology is accomplished by using the odds ratio for an arbitrary pair of rows h (e.g., blacks) and i (e.g., Hispanics), and adjacent columns j (e.g., traditional) and $j + 1$ (e.g., moderate). In comparing blacks and Hispanics with respect to gender ideology, we obtain $\hat{\tau}_3 - \hat{\tau}_1 = -.825$ or $e^{\hat{\tau}_3 - \hat{\tau}_1} = 0.438$. This can be interpreted as the odds of Hispanics having a liberal (as opposed to moderate) gender ideology, and the odds of Hispanics having a moderate (as opposed to traditional) gender ideology is about 0.438 times the odds of blacks having more liberal gender ideology.

One caution in interpretation is that if the odds ratio is less than 1, a more straightforward interpretation can be obtained by the reciprocal of the odds ratio as long as the direction of the comparison is reversed. For example, the above odds ratio (0.438) is less than 1, therefore we will use the reciprocal of 0.438, (1/0.438 = 2.282), to provide a more straightforward interpretation. By reversing the direction of comparison, the odds ratio now is blacks as opposed to Hispanics, and can be interpreted as follows: the odds of blacks holding more liberal gender ideology is about 2.282 times the odds of Hispanics holding more liberal gender ideology.

If we use blacks as a baseline category, then it is better to keep the same baseline category for other comparisons, for example, blacks versus whites. Alba (1988) recommends expressing odds ratios as uniformly above 1 or below 1, insofar as this is possible through the choice of an appropriate reference point, thus simplifying comparisons of their magnitudes. In black-white comparison, the difference $\hat{\tau}_1 - \hat{\tau}_2$ = 0.051 or 1.05 $[e^{(0.051)}]$ means that the odds of blacks having more liberal gender ideology is 1.05 times the odds of whites. Note, however, that because odds ratio of 1.0 indicates no "effect," the preceding odds ratio of 1.05 means that blacks and whites reported a similar level of gender ideology. Finally, the odds of whites holding a more liberal gender ideology is about twice the odds $[\hat{\tau}_2 - \hat{\tau}_3 = 0.774$ and $e^{(0.774)} = 2.12]$ of Hispanics. The row effect parameters show that although black and white respondents report a similar level of gender ideology, Hispanic respondents hold much more traditional attitudes toward women's roles than other groups.

In summary, the row effects model is appropriate if the study is mainly concerned with comparing levels of the nominal row variable with respect to their conditional distribution on the column variable, which is an ordinal response variable. To examine the row effects model, we first assess the fit of the model, and choose the *best fitting* model. Second, we interpret the magnitude of the association in the selected model using odds ratios.

Column Effects Model

The column effects model for an ordinal row variable and a nominal column variable is a simple variation of the row effects model. The column effects model assigns ordered scores $\{u_i\}$ to the categories of a row variable. The parameters $\{\tau_j\}$ reflects separate slopes for each column category. Suppose we use respondents' parent's social class as a row variable and respondents occupations as a column variable, then the parameters indicate the likelihood of respondents in each occupational category to come from various social class backgrounds. The column effects model that includes an association term can be expressed:

$$\log F_{ij} = \mu + \lambda_i^A + \lambda_j^B + \tau_j (u_i - \bar{u}).$$

Parallel to the interpretation of a row effects model, we evaluate the fit of a column effects model using chi-square and degrees of freedom. The odds ratios for the column effects model ($\Theta_{ij} = e^{\tau_{j+1} - \tau_j}$) are constant across different combinations of adjacent rows. Using parental social class and respondents' occupation as an example, the column effects model predicts the odds of coming from a lower (as opposed to middle) class background when a person has one occupation (say, auto mechanic) as opposed to another (say, school teacher).

Uniform Association Models

In this section we discuss a log-linear model that treats the levels of both variables as ordinal. To present this model, we use the same gender ideology variable but instead of ethnicity, the ordinal variable of religiosity is examined.

In the row effects model, we assigned scores for the ordinal column variable. Here, we use a set of integer scores for both row and column variables to reflect the ordering of these variables: $\{u_i\}$ to the rows where $u_1 < u_2 < \ldots < u_I$ and $\{v_j\}$ to the columns where $v_1 < v_2 < \ldots < v_J$. As in the previous section our goal is to pose a model that is more complex than the independence model but not saturated. This is done by including an association term that reflects the relationship between two ordinal variables. A simple log-linear model that utilizes the orderings of the row variable, religiosity (R) and column variable, gender ideology (G), is given by:

$$\log F_{ij} = \mu + \lambda_i^R + \lambda_j^G + \beta(u_i - \bar{u})(v_j - \bar{v}) \qquad (2.3)$$

To identify this model, we impose zero-sum (or ANOVA-like) constraints, thus $\sum_i \lambda_i^R = \sum_j \lambda_j^G = 0$. The transformation of scores by $(u_i - \bar{u})(v_j - \bar{v})$ corresponds to $(-1, 0, 1)$ ordering in the categories of both row and column variables. The independence model is a special case where $\beta = 0$. Goodman (1979) referred to Model 2.3 with equal-interval scores as a *uniform association model*. Model 2.3 has only one more parameter (β) being fitted than the independence model, so it has df = $IJ - [1 + (I - 1) + (J - 1)] - 1$ (for β) for testing goodness of fit.

Table 2.3 gives the observed frequencies of the cross-classification of religiosity and gender ideology for the NSFH data. Religions generally recognize and differentiate masculine and feminine principles (Stockard & Johnson, 1980). In particular, the female image in various religions is usually linked with fertility and nurturance. Because many religions define differential roles for men and women, those who consider themselves religious may tend to hold more conservative and traditional attitudes toward women's roles than those who are not at all religious. We, therefore, predict that religious respondents are more likely to hold traditional attitudes toward gender ideology. As expected, Table 2.3 shows that religious respondents seem to have a more traditional gender ideology while those who are not religious tend to be clustered in the liberal gender ideology category.

Assessing Goodness of Fit

The expected frequencies for fitting the independence and the uniform association models for religiosity and gender ideology are also given in Table 2.3. As shown, expected frequencies under the uniform association model are much closer to the observed values compared to those under the independence model, suggesting the better fit of the uniform association model. The goodness of fit of the independence model and that of the uniform association model are summarized in Table 2.4. The fit of an independence model is inadequate with $G^2 = 153.22$ based on df = 4. The uniform association model has $G^2 = 6.81$ based on df = 3, indicating a much better fit. Using a difference of chi-square test, we test the null hypothesis H_0: $\beta = 0$ to determine if

TABLE 2.3

Observed and Expected Frequencies for Independence and Uniform
Association Models of Religiosity [R] and Gender Ideology [G]

Religiosity	Gender Ideology			
	Traditional	Moderate	Liberal	Total
Religious	466	317	203	986
	(349.9)[a]	(340.8)	(295.2)	
	[449.6][b]	[339.3]	[197.2]	
Moderately Religious	135	198	183	516
	(184.6)	(179.8)	(155.7)	
	[150.5]	[188.2]	[181.3]	
Not Religious	51	120	160	331
	(117.5)	(114.4)	(99.1)	
	[51.9]	[107.6]	[171.6]	
Total	652	635	546	1,833

NOTES: a. For independence model F_{ij}.
  b. For uniform association model F_{ij}.

adding the association term β significantly improves the fit of the
independence model. The difference of G^2's is 146.41 based on df = 1.
Therefore, we reject H_0: $\beta = 0$ and conclude that adding the association
term significantly improves the fit of the model. This is strong evidence
of an association between religiosity and gender ideology.
    Alternatively, Agresti (1984) suggests that the z statistic ($\hat{\beta}/\hat{\sigma}_{\hat{\beta}}$) where
$\hat{\sigma}_{\hat{\beta}}$ is a standard error of the $\hat{\beta}$ estimates can be used for testing H_0: $\beta =$
0. Unlike the difference of G^2 tests, this statistic retains information
about direction of association, so it is useful to test the alternative hypotheses H_a: $\beta > 0$ and H_a: $\beta < 0$ with a one-tailed test. In the religiosity and

TABLE 2.4

Goodness-of-Fit Models for Religiosity and Gender Ideology Data

Model	G^2	df	prob
Independence	153.22	4	.000
Uniform Association	6.81	3	.078

gender ideology example, $(\hat{\beta}/\hat{\sigma}_{\hat{\beta}}) = (0.505/0.059) = 8.56$, thus β is statistically significant for a one-tailed z test. Therefore, we reject our null hypothesis and accept the alternative hypothesis $H_a: \beta > 0$. The positive impact means that religious respondents tend to hold more traditional gender ideology and vice versa.

Odds Ratios

We now examine the magnitude of the effect of religiosity on gender ideology. In the uniform association model (2.3), μ and the two λ parameters are the usual log-linear terms for the overall mean and the main effects of religiosity and gender ideology. In order to interpret the magnitude of β in the uniform association models, we again use log odds ratios. For an arbitrary pair of rows $a < b$ and an arbitrary pair of columns $c < d$, the log odds ratio is:

$$
\log \frac{F_{ac}F_{bd}}{F_{ad}F_{bc}}
$$

$$
= [\mu + \lambda_a + \lambda_c + \beta(u_a - \bar{u})(v_c - \bar{v}) + \mu + \lambda_b + \lambda_d + \beta(u_b - \bar{u})(v_d - \bar{v})]
$$

$$
- [\mu + \lambda_a + \lambda_d + \beta(u_a - \bar{u})(v_d - \bar{v}) + \mu + \lambda_b + \lambda_c + \beta(u_b - \bar{u})(v_c - \bar{v})]
$$

$$
= [\beta(u_a - \bar{u})(v_c - \bar{v}) + \beta(u_b - \bar{u})(v_d - \bar{v})]
$$

$$
- [\beta(u_a - \bar{u})(v_d - \bar{v}) + \beta(u_b - \bar{u})(v_c - \bar{v})]
$$

$$
= \beta(u_a v_c + u_b v_d) - \beta(u_a v_d + u_b v_c)
$$

$$
= \beta(u_b - u_a)(v_d - v_c) \tag{2.4}
$$

For rows a and b, and columns c and d in a 2×2 contingency table, the odds ratio uses four cells in a rectangular pattern. As is clear in (2.4), however, in contingency tables larger than 2×2, the log odds ratio is greater for pairs of rows and columns that are farther apart. It is useful to use the *local odds ratios* to describe properties of models for ordinal variables. Local odds ratios use cells in immediately adjacent rows and adjacent columns. Thus (2.4) can be reexpressed for the local log odds ratios as:

$$
\log \frac{F_{ij}F_{i+1,j+1}}{F_{i,j+1}F_{i+1,j}} = \beta(u_{i+1} - u_i)(v_{j+1} - v_j).
$$

$$
= \beta
$$

because $u_{i+1} - u_i = v_{j+1} - v_j = 1$ for integer scores. All local odds ratios are the same in this model. The odds ratio can also be expressed as $\Theta_{ij} = e^\beta$.

In our example, the β parameter describes the association between religiosity and gender ideology. The estimate of the association parameter is $\hat{\beta} = 0.505$. The positive value indicates that religious respondents tend to hold a more traditional gender ideology, and vice versa. As described above, uniform association models allow a simple interpretation that the odds ratio for adjacent rows (e.g., religious versus moderately religious) and adjacent columns (e.g., traditional versus moderate gender ideology) are equal. Thus the estimated uniform odds ratio, $\hat{\Theta}_{ij}$ $= e^{0.505} = 1.66$ means that for those who are not religious, the odds of holding a liberal (as opposed to moderate), and a moderate (as opposed to traditional) gender ideology is about 1.66 times the odds of those who are moderately religious. Likewise, the odds of moderately religious persons holding more liberal attitudes toward men's and women's family roles and employment is about 1.66 times the odds of those who are very religious.

Another useful interpretation compares the difference between the lowest and highest scores of religiosity (very religious versus not at all religious) and gender ideology (traditional versus liberal). The odds ratio of not at all religious (as opposed to very religious) respondents holding a liberal instead of traditional gender ideology can be estimated by the following:

$$\hat{\Theta}_{ij} = e^{[\hat{\beta}(u_3 - u_1)(v_3 - v_1)]} = e^{[0.505(3-1)(3-1)]} = 7.54$$

Therefore, for non-religious respondents, the odds of having a more egalitarian gender ideology is 7.54 times the odds of very religious respondents having a more egalitarian ideology.

Using the information we have, it is also possible to generate a confidence interval for the odds ratio. For example, a 95% confidence interval for the estimated odds ratio to be equal to the population odds ratio, $\Theta_{ij} = e^\beta$, can be computed by:

$$\hat{\Theta}_{ij} = e^{[\hat{\beta} \pm z_{.025}\hat{\sigma}_\beta]}$$

where $\hat{\beta}$ is an estimated association parameter, $z_{.025}$ is a z-value that cuts off 2.5% from the upper tail (and by symmetry, also 2.5% from the lower tail), and $\hat{\sigma}_\beta$ is a standard error of the β term. Computing this for our example, we obtain $e^{(0.505 \pm 1.95 \times 0.059)}$, or (1.48, 1.86). That is, we can be 95% confident that the true odds ratio is between 1.48 and 1.86.

Assignment of Scores

As previously demonstrated, the log-linear models for ordinal variables require the assignment of scores to reflect the ordinality of the variables. Parameter interpretations are simplest for equally spaced scores. It is frequently unclear, however, how to assign scores, and further, we may not want to assume equal spacing. Typical categories of marital happiness, for example, include "very happy," "pretty happy," and "not too happy." Although "very happy" is more than "pretty happy," the actual distance between the two response categories is not as clearly defined as interval measures. For example, Clogg (1982a) has shown that the distances between the categories on the happiness scale are not equal. He demonstrated that the distance between "not too happy" and "pretty happy" categories is about three times greater than the distance between "pretty happy" and "very happy" responses.

There is no simple answer concerning how sensitive ordinal analyses are to the choice of scores. Studies that use a simulation technique have generally concluded that assumption of equal distance for ordinal variables results in little error (O'Brien, 1979). But others (e.g., Graubard & Korn, 1987) show examples in which different scoring systems give quite different results. When there is no natural set of scores, Agresti (1990) suggests that it is wise to conduct a sensitivity analysis. Such an analysis involves assigning scores in a variety of ways that seem reasonable, and checking whether substantive conclusions depend on the choice.

The necessity of assigning scores for ordinal classifications is not always a draw back. For example, if the uniform association model fits well with unit row and column scores, what we have is a simple and parsimonious description of the nature of the association given by uniform local odds ratios. This simplicity can be viewed as a positive aspect of scoring, regardless of whether the scores are reasonable indexes of "true" distances between ordered categories (Agresti, 1990). The perceived advantage of procedures not requiring preassigned scores was also pointed out to be illusory (Graubard & Korn, 1987). It was noted that procedures that automatically generate scores, such as non-parametric methods that utilize mid-ranks, may produce inappropriate scores. There are models, however, that can be used when the assignment of unit-based scores is considered inappropriate. The following section presents these models.

Row and Column Effects Model
(The RC Model)

As previously demonstrated, odds ratios and the goodness of fit of the models largely depend on the choice of scoring system. This section considers a model that can be applied to ordinal variables but does not require the assignment of scores. This model, known as the *Row and Column Effects Model*, is a special case of a log-multiplicative model in which row and column scores are parameters. The models are called log-multiplicative because the log of expected frequency, $\log F_{ij}$, is a multiplicative function of the model parameters instead of linear in the natural logs.

The uniform association model is a special case of the row effects model and the column effects model. The row effects model has parameter row scores, and the column effects model has parameter column scores. These models, in turn, are special cases of the row and column effects model that replace both row and column scores by parameters. Let us illustrate the row and column effects model for two-way tables using the same religiosity and gender ideology example. When we replace the row scores $\{u_i\}$ and column scores $\{v_j\}$ in the uniform association model by parameters $\{\mu_i\}$ and $\{v_j\}$, respectively, we obtain the row and column effects model. The two-dimensional row and column effects model is expressed by:

$$\log F_{ij} = \mu + \lambda_i^R + \lambda_j^G + \beta\mu_i v_j \qquad (2.5)$$

where $\sum_i \lambda_i^R = \sum_j \lambda_j^G = \sum_i \mu_i = \sum_j v_j = 0$. Model 2.5 resembles the row effects Model 2.1 when we treat the $\{\mu_i\}$ as row effects and the $\{v_j\}$ as scores. Generally, the $\{\mu_i\}$ in (2.5) can be regarded as row effects and the $\{v_j\}$ can be regarded as column effects. This model is frequently referred to as the *RC model* to reflect its multiplicative row and column effects.

The RC model is different from the row effects and the column effects models because it is invariant to interchanges of rows and columns. For example, if Model 2.5 fits reasonably well, interchange of rows and columns will not affect the fit of the model and will also demonstrate a reasonable fit. According to Agresti (1984), if this model fits well and produces parameter score estimates that have an ordinal trend, then the uniform association models would also fit well if the scores assigned for that model had similar spacings.

Fitting this model to the data in Table 2.3 yields G^2 of .41 based on df = 1 indicating an extremely good fit. As shown before, the uniform association model also fits the data but not as well as the RC model. The difference of G^2's tests the hypothesis that row and column scores in the RC model are the same as those in uniform association model (i.e., equal-interval). The statistic, G^2(uniform association model) − G^2(RC model) = 6.4 with df = 2 indicates that the parameter scores given in the RC model give a statistically better fit than equal-interval scores.

Odds Ratios

The invariance to the orderings of categories implies that variables are treated as nominal by this model. However, the model can still be used to assess the ordinal characteristics of the data. This can be accomplished by examining local odds ratios:

$$\log \frac{F_{ij} F_{i+1,j+1}}{F_{i,j+1} F_{i+1,j}} = (\mu + \lambda_i^A + \lambda_j^B + \beta\mu_i v_j + \mu + \lambda_{i+1}^A + \lambda_{j+1}^B + \beta\mu_{i+1} v_{j+1})$$

$$- (\mu + \lambda_i^A + \lambda_{j+1}^B + \beta\mu_i v_{j+1} + \mu + \lambda_{i+1}^A + \lambda_j^B + \beta\mu_{i+1} v_j)$$

$$= \beta(\mu_i v_j + \mu_{i+1} v_{j+1}) - \beta(\mu_i v_{j+1} + \mu_{i+1} v_j)$$

$$= \beta(\mu_{i+1} - \mu_i)(v_{j+1} - v_j) \tag{2.6}$$

If these local odds ratios indicate ordinality in the scores, then all local associations have the same sign. Lack of ordinality in the scores means there exists no monotonic association. This indicates that local associations are positive in some locations but negative in other locations. Because of the constraints imposed on the model parameters, $I - 2$ of the $\{\mu_i\}$ and $J - 2$ of the $\{v_j\}$ are linearly independent. Therefore df is $IJ - [1 + (I - 1) + (J - 1) + 1 + (I - 2) + (J - 2)] = (I - 2)(J - 2)$ for goodness-of-fit tests.

Let us consider the estimated scores for the RC model with religiosity and gender ideology variables (2.5). We get $\hat{\mu}_1 = 0.97$, $\hat{\mu}_2 = 1.06$, $\hat{\mu}_3 = $

1.01 for the levels of religiosity and $\hat{v}_1 = 0.55$, $\hat{v}_2 = 1.03$, $\hat{v}_3 = 1.44$ for gender ideology. The $\{v_j\}$ are monotonic and nearly evenly spaced. Thus the use of unit-spaced scores for the gender ideology variable seems appropriate. The $\{\mu_i\}$, on the other hand, are not monotonic suggesting that the use of preassigned scores may be problematic for a religiosity variable.

The estimated local odds ratios are positive: 2.03 (traditional versus moderate gender ideology given religious versus moderately religious responses), 1.53 (moderate versus liberal gender ideology given religious versus moderately religious responses), 1.77 (traditional versus moderate gender ideology given moderately versus not at all religious responses), and 1.34 (moderate versus liberal gender ideology given moderately versus not at all religious responses). These odds ratios indicate a tendency for less religious respondents to report a more liberal gender ideology. More specifically, the odds ratio of 2.03 means that the odds of respondents who are moderately religious holding a moderate level of gender ideology as opposed to traditional ideology is about twice the odds of very religious respondents reporting the same level of gender ideology. Other odds ratios can be similarly interpreted.

Unlike the uniform association, row effects, and column effects models, the RC model is not log-linear, because the log expected frequency is a multiplicative (rather than linear) function of the model parameters μ_i and v_j. These models have been extensively discussed by Anderson (1980), Clogg (1982a, 1982b), and Goodman (1979, 1981a, 1981b).

Because of an arbitrary nature of assigning scores to an ordinal variable, replacing fixed scores by parameters is an attractive alternative for some applications. The RC model, however, is not necessarily *better* than log-linear models, because it presents complications that do not exist in log-linear modeling. Agresti (1990) pointed out that the likelihood of the RC model may not be concave, and it may have local maximum. This means that there is a value of the fitting function in the RC models that appears to be the largest possible value to reproduce the observed data, when actually there are other larger values. It is also problematic to compare the independence and the RC models. Haberman (1981) demonstrates that the difference of G^2's between independence and the RC models has a distribution that is not chi-squared. Therefore, the use of chi-squared distribution as a test statistic may not be justified.

In summary, the RC model may generate similar findings as the uniform association model. The RC model, however, may be useful when the distance between scores is unknown, and to test the validity of equal-interval scores assigned to rows and columns. Researchers, however, need to be aware of the strengths and weaknesses of the RC models.

Odds Ratios for Two-Way Tables:
Summary

A basic question researchers usually pose when analyzing ordinal data is "Does Y tend to increase as X increases?" Bivariate analyses of interval-scale variables often summarize covariation by the Pearson correlation, which describes the degree to which Y has a linear relationship with X. Ordinal variables do not have a defined metric, so the notion of linearity is not meaningful. However, the inherent ordering of categories allows consideration of monotonicity—for instance, whether Y tends to increase as X does. Each odds ratio, Θ_{ij}, describes such an association present in a particular region of the $I \times J$ table.

The odds ratio can equal any nonnegative number, with the ratio of 1.0 indicating no association. Values of odds ratios that are farther from 1.0 in a given direction represent stronger levels of association. The odds ratio does not change value when the orientation of the table is reversed so that the rows become the columns and the columns become the rows. Therefore, it is not necessary to identify one classification as the response variable in order to calculate Θ_{ij}. Another property of Θ_{ij} is that its value does not change if both cell frequencies within any row are multiplied by a nonzero constant, or if both cell frequencies within any column are multiplied by a constant. Hence Θ_{ij} estimates the same characteristics even if disproportionately large or small samples are selected from the various marginal categories of a variable. In addition, if the order of the rows is reversed or if the order of the columns is reversed, the new value of Θ_{ij} is simply the inverse of the original value.

In Table 2.5, odds ratios (Θ_{ij}) of various models for the 3×3 religiosity and gender ideology table (see Table 2.3) are summarized. In the two-way *saturated model*, odds ratios are identical to the observed cross-product ratio, $(f_{ij}f_{i+1,j+1})/(f_{i,j+1}f_{i+1,j})$. The first odds ratio of 2.16 for the saturated model means that the odds of respondents who are moderately religious indicating a moderate (as opposed to traditional) level of gender ideology is 2.16 times the odds of religious respondents holding more liberal ideology. All the local odds ratios of *independence model* are 1, indicating no association between two vari-

TABLE 2.5

Local Odds Ratios for 3 × 3 Table of Religiosity and Gender Ideology

				Model			
Rows	Columns	Saturated	Independence	Row Effects	Column Effects	Uniform Association	Row-Column (RC)
1, 2	1, 2	2.16	1.00	1.78	1.95	1.66	2.03
1, 2	2, 3	1.44	1.00	1.78	1.44	1.66	1.53
2, 3	1, 2	1.60	1.00	1.48	1.95	1.66	1.77
2, 3	2, 3	1.44	1.00	1.48	1.44	1.66	1.34
		Θ_{ij}	1	$\Theta_{i.}$	$\Theta_{.j}$	Θ	$\Theta_i \Theta_j$

ables. In the *row effects model* the odds ratios vary across rows but are invariant across column categories ($\Theta_{i.}$). The subscripted dots in this table denote that the particular parameters do not depend on the value of the subscripts that they replace. Therefore, $\Theta_{i.}$ is a quantity that does not depend on the subscript j (column level). That is, the row effects model assumes that within the adjacent rows (religious versus moderately religious) the odds of adjacent columns (traditional versus moderate, and moderate versus liberal gender ideology) are identical. Conversely, the odds ratios in the *column effects model* vary only across columns ($\Theta_{.j}$). In the *uniform association model* all Θ_{ij} are equal to a constant Θ that describes the association between two variables. Therefore, the local odds ratio to describe adjacent rows and adjacent columns is constant and is the same for any arbitrary pair of rows and an arbitrary pair of columns.

The local odds ratios in the *multiplicative row and column effects model* (RC model) vary across rows and columns ($\Theta_i \Theta_j$). In this table, all the odds ratios are greater than 1.0, indicating the monotonic associations. If local associations are greater than 1.0 in some locations and less than 1.0 in other locations, the model lacks monotonic associations. That is why this model can be used to detect and describe ordinal characteristics of data although the variables are treated as nominal.

In summary, a single number such as the odds ratio conveniently describes various associations in the table. As shown in Table 2.5, among ordinal models, the uniform association model yields the simplest interpretation of odds ratios because the model's association in the table can be described by a constant odds ratio. The independence model is often substantively less important. On the other hand, the uniform association model provides the opportunity to test the hypothesized model with a straightforward interpretation of odds ratios. Both row effects and column effects models also provide simple interpretations of odds ratios. Because one of the objectives of log-linear modeling is to find a parsimonious model with a straightforward interpretation of parameters, uniform association, row effects, and column effects models provide alternative choices that were unavailable in log-linear models for nominal variables.

Ordinal Log-Linear Models for Higher Ordered Tables

The row (or column) effects and uniform association models can be generalized without difficulty to multidimensional tables having at least one ordinal variable. But for ease of exposition, ordinal log-linear models for a three-way table are considered in this section. The *nominal* hierarchical log-linear models for three-way tables take various forms. These models range from a simple independence model to a model that contains all the two-factor partial associations but not the three-factor interaction term (see Table 1.4). The most complex nominal model with three variables is a saturated one that has the three-factor interaction term. If one or more of the variables in the cross-classification is ordinal, however, there exists a nonsaturated model with a three-factor interaction term. We will begin with the models that include both nominal and ordinal variables. Log-linear models for higher ordered tables with all ordinal variables will then be presented.

Models for Nominal-Ordinal Variables

To illustrate ordinal log-linear models for three-way cross-classification, we first examine a model that includes both nominal and ordinal variables. Table 2.6 is a $3 \times 3 \times 3$ cross classification of ethnicity (E), religiosity (R), and gender ideology (G) for a sample of married respondents in NSFH data. In the previous analyses, we found that both ethnicity and religiosity are significantly related to one's attitudes toward women's family roles and employment. More specifically, we found that blacks hold the most liberal and Hispanics hold the most traditional gender ideology. With respect to religiosity, there was a general pattern that more religious respondents were likely to hold a more traditional gender ideology. Although these bivariate relationships

TABLE 2.6

Observed and Expected Frequencies for Partial Association Model
for Ethnicity [E], Religiosity [R], and Gender Ideology [G] Data

		Gender Ideology			
Ethnicity	Religiosity	Traditional	Moderate	Liberal	Total
Black	Religious	58 (55.5)[a]	45 (56.0)	49 (42.3)	152
Black	Moderately Religious	11 (9.6)	17 (16.0)	21 (19.9)	49
Black	Not Religious	3 (1.7)	4 (4.6)	7 (9.5)	14
White	Religious	317 (309.5)	242 (245.2)	145 (145.3)	704
White	Moderately Religious	105 (116.3)	157 (152.4)	148 (149.3)	410
White	Not Religious	41 (44.4)	109 (96.1)	150 (155.6)	300
Hispanic	Religious	83 (77.3)	24 (30.8)	8 (9.2)	115
Hispanic	Moderately Religious	16 (21.0)	17 (13.8)	13 (6.8)	46
Hispanic	Not Religious	7 (5.8)	6 (6.3)	2 (5.1)	15
Total		641	621	543	1,805

NOTE: a. For partial association model F_{ij}.

were found in the previous analyses, we have not simultaneously examined the relationships among the three variables. For example, we have not addressed such questions as "Are religious blacks more likely to hold a traditional gender ideology than blacks who are not at all religious? How about whites and Hispanics?" and "Is the impact of ethnicity on gender ideology greater than that of religiosity?"

As shown in Table 2.6, regardless of race there is a trend for religious respondents to hold a more traditional ideology. However, it is not clear if less religious respondents hold a more liberal ideology. These relationships will be further illuminated by ordinal log-linear models.

For a nominal ethnicity variable, and ordinal religiosity and gender ideology variables, a basic *partial association model* that includes two-factor but not three-factor association terms can be expressed as follows:

$$\log F_{ijk} = \mu + \lambda_i^E + \lambda_j^R + \lambda_k^G$$

$$+ \tau_i^{ER} (v_j - \overline{v}) + \tau_i^{EG}(w_k - \overline{w})$$

$$+ \beta^{RG} (v_j - \overline{v})(w_k - \overline{w}) \qquad (2.7)$$

where usual constraints, $\sum_i \lambda_i^E = \sum_j \lambda_j^R = \sum_k \lambda_k^G = \sum_i \tau_i^{ER} = \sum_i \tau_i^{EG} = 0$, are imposed. In Model 2.7, the association term between religiosity and gender ideology, $\beta^{RG} (v_j - \overline{v})(w_k - \overline{w})$, has the same form as the association term for the uniform association Model 2.3. The association terms $\tau_i^{ER}(v_j - \overline{v})$ and $\tau_i^{EG}(w_k - \overline{w})$ for the pairs of nominal and ordinal variables have the same form as the association terms for the row effects Model 2.1. The expected frequencies of this partial association model are also given in Table 2.6.

Assessing Goodness of Fit

In addition to the partial association models, a variety of more parsimonious association models can be considered with a three-way table. Table 2.7 summarizes the results of goodness-of-fit tests for various ordinal log-linear models for a three-way table. The independence model, [E][R][G], is inadequate with $G^2 = 288.46$ based on df = 20. Although the fit of the model is poor, the ordinal models with at least 1 two-factor association term seem to have a much better fit (see Models 2, 3, and 4) than the independence model. The model with partial association terms, [ER][EG][RG], fits the best with $G^2 = 20.44$ based on df = 15.

Table 2.8 shows the results of difference of G^2 tests for various models. Comparing Models 5, 6, and 7 each with 8, by the G^2 and df differences, we find that adding [ER] or [EG] or [RG] association terms significantly improves the fit of the model. The G^2 difference between Models 5 and 8 testing the null hypothesis $\tau_i^{ER} = 0$ is 53.75 with a difference in df 2, thus we reject the hypothesis. Likewise, the G^2 difference between Models 6 and 8 is 49.36 based on df = 2. Rejecting $H_0: \tau_i^{EG} = 0$ means that there is a significant difference across three

TABLE 2.7

Goodness of Fit of Independence and Association Models for
Ethnicity [E], Religiosity [R], and Gender Ideology [G] Data

Model	Fitted Marginals	G^2	df	prob
1	[E][R][G]	288.46	20	.000
2	[E][RG]	129.40	19	.000
3	[R][EG]	233.25	18	.000
4	[G][ER]	228.87	18	.000
5	[EG][RG]	74.19	17	.001
6	[ER][RG]	69.80	17	.001
7	[ER][EG]	173.66	16	.000
8	[ER][EG][RG]	20.44	15	.156

ethnic groups in terms of gender ideology, holding religiosity constant.
The level of religiosity is also significantly related to the gender ideology as we reject H_0: $\beta^{RG} = 0$, holding ethnicity constant.

Finally, if we want to address a question of whether religious blacks
(or whites or Hispanics) hold a more traditional gender ideology than
nonreligious blacks, we need to include the three-factor interaction
term, [ERG]. The model with this term yields $G^2 = 18.82$ based on df =
13. Although the fit of the model is reasonable, the [ERG] association
term is not significant because the difference in G^2's between this model
and the partial association model is 1.62 with df of 2. Substantively, this
means that the relationship between gender ideology and ethnicity is
independent of one's religiosity. Similarly, the relationship between
ideology and religiosity is not influenced by one's ethnicity.

Based on these results and substantive considerations, we choose
Model 8, [ER][EG][RG], as the best fitting model in this example.
According to this model, ethnicity and gender ideology have a uniform

TABLE 2.8

Comparison Among Models Fitted in Table 2.7

Association Terms	Models Compared	Difference in G^2	Difference in df	prob	
[ER]	5 and 8	53.75	2	.000	sig
[EG]	6 and 8	49.36	2	.000	sig
[RG]	7 and 8	153.22	1	.000	sig

association that is the same for each level of religiosity. Alternatively, religiosity and gender ideology have a uniform association that is the same across ethnicity.

Odds Ratios

The next step is to assess the magnitudes of the effect of ethnicity and religiosity on gender ideology in the selected partial association model. To interpret these effects, we use the odds ratios $\Theta_{ij(k)}$, $\Theta_{i(j)k}$, and $\Theta_{(i)jk}$ that describe the local conditional associations between two variables within a fixed level of the third variable. For the partial association model with nominal-ordinal variables, the local log odds ratios, log $\Theta_{ij(k)}$, are:

$$\log \frac{F_{ijk} F_{i+1,j+1,k}}{F_{i,j+1,k} F_{i+1,j,k}}$$

$$= [\tau_i^{ER}(v_j - \bar{v}) + \tau_{i+1}^{ER}(v_{j+1} - \bar{v})] - [\tau_i^{ER}(v_{j+1} - \bar{v}) + \tau_{i+1}^{ER}(v_j - \bar{v})]$$

$$= \tau_{i+1}^{ER}(v_{j+1} - v_j) - \tau_i^{ER}(v_{j+1} - v_j)$$

$$= \tau_{i+1}^{ER} - \tau_i^{ER} \tag{2.8}$$

because $v_{j+1} - v_j = 1$. Other log odds ratios are similarly derived:

$$\log \Theta_{i(j)k} = \tau_{i+1}^{EG} - \tau_i^{EG},$$

and

$$\log \Theta_{(i)jk} = \beta^{RG}.$$

The τ_i^{ER}, τ_i^{EG}, and β^{RG} parameters describe the pairwise partial association between each pair of variables, and are the same for all levels of the third variable. Therefore, the τ_i^{ER} represent row effects of ethnicity on the ethnicity-religiosity association that are homogenous across levels of gender ideology. Similarly, the τ_i^{EG} represent row effects of ethnicity on the ethnicity-gender ideology association that are homogeneous across levels of religiosity. The β^{RG} parameter refers to the uniform association between religiosity and gender ideology, which is homogeneous across ethnicity.

TABLE 2.9

Odds for a Selected Partial Association Model [ER][EG][RG]

Parameter[a]	Description	τ or β	e^τ or e^β
τ_{1j}^{ER}	The odds of not being religious (as opposed to moderately religious), or the odds of being moderately religious (as opposed to religious), for blacks.	−0.412	0.662*
τ_{2j}^{ER}	The odds of not being religious (as opposed to moderately religious), or the odds of being moderately religious (as opposed to religious), for whites.	0.368	1.445*
τ_{3j}^{ER}	The odds of not being religious (as opposed to moderately religious), or the odds of being moderately religious (as opposed to religious), for Hispanics.	0.044	1.045
τ_{1k}^{EG}	The odds of holding liberal (as opposed to moderately liberal/traditional), or the odds of holding moderately liberal/traditional (as opposed to traditional) attitudes toward women's roles, for blacks.	0.391	1.478*
τ_{2k}^{EG}	The odds of holding liberal (as opposed to moderately liberal/traditional), or the odds of holding moderately liberal/traditional (as opposed to traditional) attitudes toward women's roles, for whites.	0.148	1.160*
τ_{3k}^{EG}	The odds of holding liberal (as opposed to moderately liberal/traditional), or the odds of holding moderately liberal/traditional (as opposed to traditional) attitudes toward women's roles, for Hispanics.	−0.539	0.583*
β^{RG}	The odds of holding liberal (as opposed to moderately liberal/traditional), or the odds of holding moderately liberal/traditional (as opposed to traditional) attitudes toward women's roles, for those with lower level of religiosity.	0.711	2.036*

NOTE: a. (i) refers to categories of ethnicity, (j) refers to categories of religiosity, and (k) refers to categories of gender ideology.
*$p < .05$

The association terms in this model can be described in several ways. First, we can interpret individual parameters. Table 2.9 lists the parameter estimates and interpretation of effects of the partial association model (2.7). As shown, the individual τ parameters support some of the previous findings: Blacks are more likely to be religious while whites are least likely to be religious. The trend for Hispanics is not clear. Blacks are more likely to hold liberal attitudes toward women's roles while Hispanics are reported to hold a more traditional gender ideology. Less religious respondents were reported to have a liberal gender ideology. Second, various odds ratios can be used to describe the association terms. For example, black-white comparison can be made by computing odds ratios: $\tau_{1j}^{ER} - \tau_{2j}^{ER} = -0.412 - 0.368 = -0.780$ and $e^{(\tau_{1j}^{ER} - \tau_{2j}^{ER})} = e^{-0.780} = 0.458$. This means that the odds of blacks being religious are almost twice ($2.181 =$ inverse of 0.458) the odds of whites being religious. Likewise, $e^{(\tau_{2j}^{ER} - \tau_{3j}^{ER})} = 1.383$ means that the odds of whites being religious are 1.383 times the odds of Hispanics being religious. The association terms between ethnicity and gender ideology show that the odds of blacks holding liberal attitudes toward women's roles are 1.275 times $[e^{(\tau_{1k}^{EG} - \tau_{2k}^{EG})}]$ the odds of whites holding similar attitudes. On the other hand, the odds of whites holding more liberal ideology are 1.988 times $[e^{(\tau_{2k}^{EG} - \tau_{3k}^{EG})}]$ the odds of Hispanics holding more liberal ideology. Our previous findings in two-way tables are again supported in this model with three variables.

In addition, the parameter estimate for the religiosity-gender ideology association ($\hat{\beta}^{RG}$) indicates that the odds of less religious people reporting a liberal gender ideology are 2.03 times the odds of more religious people reporting a similar level of ideology. It is important to note that there is only one coefficient to describe this effect. Therefore, the effect of going up one level higher on religiosity on the odds of moving up in gender ideology is the same regardless of a person's ethnicity.

Models for Ordinal Variables

We now consider the partial association and interaction models in which all three variables are ordinal. To illustrate this model, we use the same ordinal variable, gender ideology, but this time we look at husband's and wife's ideology separately, and consider gender ideology as an antecedent variable. We will then use another ordinal variable, the extent of couple's sharing of housework including preparing meals, washing dishes and meal cleanup, shopping, washing and ironing clothes, and cleaning house.

Scholars studying household labor allocation frequently hypothesized that both husband's and wife's gender ideology influence the way couples share routine housework (Ishii-Kuntz & Coltrane, 1992; Kamo, 1988). This approach assumes that people are socialized to adopt values and beliefs about the appropriateness of various tasks for men and women, and that such values encourage or inhibit the sharing of domestic labor. Some researchers report that men's ideology is more influential in shaping divisions of labor, whereas others report that women's ideology is the most important.

As shown in Table 2.10, more than half of the respondents reported "sometimes" sharing the housework. To model the associations among husband's gender ideology (H), wife's gender ideology (W), and sharing of housework (S), we will assign scores $\{u_i\}$, $\{v_j\}$, and $\{w_k\}$ to the levels of H, W, and S, respectively. A simple generalization from the two-way uniform association model (2.3) for a partial association model in a three-way table is:

$$\log F_{ijk} = \mu + \lambda_i^H + \lambda_j^W + \lambda_k^S$$

$$+ \beta^{HW}(u_i - \overline{u})(v_j - \overline{v}) + \beta^{HS}(u_i - \overline{u})(w_k - \overline{w}) + \beta^{WS}(v_j - \overline{v})(w_k - \overline{w})$$

$$(2.9)$$

The β^{HW}, β^{HS}, and β^{WS} parameters describe the partial associations. Model 2.9 is referred to as a *homogenous uniform association model* (Agresti, 1984, 1990). This model says that the association in each table is uniform and that this uniform association is homogeneous across groups. This model has only three more parameters than the independence model, so df = $IJK - I - J - K - 1$, and the model is always unsaturated.

The ordinal log-linear models also allow the unsaturated model with the three-factor interaction terms. A simple interaction model for the above example is

$$\log F_{ijk} = \mu + \lambda_i^H + \lambda_j^W + \lambda_k^S$$

$$+ \beta^{HW}(u_i - \overline{u})(v_j - \overline{v}) + \beta^{HS}(u_i - \overline{u})(w_k - \overline{w}) + \beta^{WS}(v_j - \overline{v})(w_k - \overline{w})$$

$$+ \beta^{HWS}(u_i - \overline{u})(v_j - \overline{v})(w_k - \overline{w}) \qquad (2.10)$$

This model has only one more parameter than Model 2.9, so df = $IJK - I - J - K - 2$, and the model is unsaturated whenever I, J, or K exceeds 2. This model is called the *uniform interaction model* (Goodman, 1979).

Interaction models can also be formed by using ordinal associations for some (as opposed to all) pairs of variables. For example, we can test

TABLE 2.10
Observed and Expected Frequencies for Interaction Model for Husband's [H] and Wife's [W] Gender Ideology and Housework [S] Data

Husband's Gender Ideology	Wife's Gender Ideology	Sharing of Housework			
		Rarely	Sometimes	Often	Total
Traditional	Traditional	73 (69.8)[a]	137 (134.3)	43 (37.1)	253
Traditional	Moderate	38 (38.4)	89 (103.0)	32 (39.6)	159
Traditional	Liberal	13 (14.4)	59 (53.7)	35 (28.8)	107
Moderate	Traditional	21 (28.3)	69 (73.0)	28 (27.0)	118
Moderate	Moderate	36 (28.8)	121 (105.0)	49 (55.0)	206
Moderate	Liberal	17 (19.9)	103 (102.9)	72 (76.3)	192
Liberal	Traditional	8 (5.9)	21 (20.5)	6 (10.2)	35
Liberal	Moderate	8 (11.2)	56 (55.5)	47 (39.5)	111
Liberal	Liberal	17 (14.3)	95 (102.2)	106 (104.7)	218
Total		231	750	418	1,399

NOTE: a. For interaction model F_{ij}.

a model with only the three-factor interaction specified with ordinal format, thus:

$$\log F_{ijk} = \mu + \lambda_i^H + \lambda_j^W + \lambda_k^S + \lambda_{ij}^{HW} + \lambda_{ik}^{HS} + \lambda_{jk}^{WS}$$

$$+ \beta^{HWS}(u_i - \overline{u})(v_j - \overline{v})(w_k - \overline{w})$$

TABLE 2.11

Goodness of Fit of Independence and Association Models
for Husband's [H] and Wife's [W] Gender Ideology,
and Sharing of Housework [S] Data

Model	Fitted Marginals	G^2	df	prob
1	[HW][HS]	58.80	18	.000
2	[HW][WS]	49.48	18	.000
3	[HS][WS]	199.18	18	.000
4	[HW][HS][WS]	20.28	17	.260
5	[HWS]	20.23	16	.210

This model has df $= (I - 1)(J - 1)(K - 1) - 1$, so unlike the interaction model for nominal variables, it is unsaturated whenever I, J, or K exceeds 2. There is constant value for log odds ratios when this model is applied with integer scores. This model, however, does not have uniform association within the partial tables.

Assessing Goodness of Fit. In Table 2.11, goodness of fit of several association models is presented. The fit of Models 1, 2, and 3 with a combination of two-factor association terms are relatively poor. The homogeneous uniform association model with all the two-factor association terms, [HW][HS][WS], fits well with $G^2 = 20.28$ based on df = 17. The uniform interaction model that includes the [HWS] association also provides an acceptable fit ($G^2 = 20.23$, df = 16). The difference of chi-square test between the above two models, however, indicates that the [HWS] term does not significantly improve the fit of the model ($\Delta G^2 = .05$ with Δdf = 1).

If we follow statistical criteria strictly and try to avoid a Type I error, we will reject the uniform interaction model as not significantly better than the partial association model and hence accept the hypothesis of no three-factor interaction effects. However, we seem to have encountered a gray area in which our decisions may be influenced as much by the substantive and theoretical aims that motivate the research as by strict statistical reasoning. Although the partial association model gives a satisfactory fit to the full cross-tabulation and is more parsimonious, we may wish to accept the uniform interaction model because the inclusion of the three-factor interaction term makes more substantive and theoretical sense. On the other hand, regardless of what the theory says, if the finding is very nonsignificant, as it is here, then it is not supported empirically and probably should be dropped from the model, particularly if one is trying to come up with a final parsimonious model. In the following section, we will interpret the odds ratios for both homogeneous uniform association and uniform interaction models for pedagogical purposes.

Odds Ratios: Uniform Association Model

In homogeneous uniform association Model 2.9, there is no third-order interaction term. By using equal-interval scores for all three variables, the local conditional log odds ratio in the partial association model, $\log \Theta_{ij(k)}$, simplifies to

$$\log \frac{F_{ijk}F_{i+1,j+1,k}}{F_{i,j+1,k}F_{i+1,j,k}} = [\beta^{HW}(u_i - \bar{u})(v_j - \bar{v}) + \beta^{HW}(u_{i+1} - \bar{u})(v_{j+1} - \bar{v})]$$

$$- [\beta^{HW}(u_i - \bar{u})(v_{j+1} - \bar{v}) + \beta^{HW}(u_{i+1} - \bar{u})(v_j - \bar{v})]$$

$$= \beta^{HW}(u_{i+1} - u_i)(v_{j+1} - v_j)$$

$$= \beta^{HW} \tag{2.11}$$

Likewise, we derive $\log \Theta_{i(j)k} = \beta^{HS}$, and $\log \Theta_{(i)jk} = \beta^{WS}$. According to these, the local odds ratio is uniform for each pair of variables, and the strength of association is homogeneous across the levels of the third variable. The interpretation of association is a simple extension of two-way association models.

The estimates of HW, HS, and WS parameters under the uniform association model (2.9) are $\hat{\beta}^{HW} = 0.895$, $\hat{\beta}^{HS} = 0.438$, and $\hat{\beta}^{WS} = 0.488$. These association parameter estimates indicate that the odds of husbands with a liberal gender ideology having wives whose ideology is also liberal are 2.45 ($e^{0.895}$) times the odds of husbands with less liberal gender ideology. The odds of husbands with a nontraditional gender ideology to share housework more frequently with their wives is about 1.54 ($e^{0.438}$) times the odds of husbands with a traditional gender ideology. Likewise, the odds of wives with a more liberal gender ideology sharing housework with their husbands are 1.63 ($e^{0.488}$) times the odds of wives with a more traditional ideology.

Odds Ratios: Uniform Interaction Model

The local three-factor interaction of the uniform interaction model (2.10) is also expressed by log odds ratio of odds ratio for unit-spaced scores, $\log \Theta_{ijk}$:

$$\log \frac{\Theta_{ij(k+1)}}{\Theta_{ij(k)}} = \log \Theta_{ij(k+1)} - \log \Theta_{ijk}$$

$$= \beta^{HWS}(u_{i+1} - u_i)(v_{j+1} - v_j)(w_{k+1} - w_k)$$

$$= \beta^{HWS} \tag{2.12}$$

Also,

$$\Theta_{ijk} = \frac{\Theta_{ij(k+1)}}{\Theta_{ij(k)}} = \frac{\Theta_{i(j+1)k}}{\Theta_{i(j)k}} = \frac{\Theta_{(i+1)jk}}{\Theta_{(i)jk}} = \beta^{HWS}$$

Essentially, Θ_{ijk} describes the local interaction in a $2 \times 2 \times 2$ section of the table consisting of adjacent rows, adjacent columns, and adjacent layers. There is an absence of three-factor interaction if all $(I-1)(J-1)(K-1)$ of the Θ_{ijk} equal 1.0. The model's interpretation is that a log odds ratio for any two variables changes linearly across the levels of the third variable. The local interaction is thus constant and equals β^{HWS} for all $2 \times 2 \times 2$ subtables formed from adjacent rows, adjacent columns, and adjacent layers.

The estimate for the three-factor association of the uniform interaction model, $\log \Theta_{ijk}$, is $\hat{\beta} = 0.032$. Any local odds ratio in the relationship between wife's gender ideology and sharing housework for households with more liberal husbands is estimated to be 1.03 times ($e^{0.032}$) higher than the corresponding local odds ratio for more traditional husbands. This indicates that the relationship between a wife's gender ideology and the sharing of housework is not significantly influenced by a husband's ideology. Likewise, the relationship between a husband's ideology and sharing of housework does not depend on a wife's gender ideology. This was statistically apparent from the comparison of fits between partial association and interaction models. However, this may be a substantively important finding that cannot be statistically examined with the partial association model.

For two-factor association terms in the uniform interaction model (2.10), the local log odds ratio, $\log \Theta_{ij(k)}$, can be expressed:

$$\log\frac{F_{ijk}F_{i+1,j+1,k}}{F_{i,j+1,k}F_{i+1,j,k}}$$

$$= [\beta^{HW}(u_i-\overline{u})(v_j-\overline{v}) + \beta^{HWS}(u_i-\overline{u})(v_j-\overline{v})(w_k-\overline{w})$$

$$+ \beta^{HW}(u_{i+1}-\overline{u})(v_{j+1}-\overline{v}) + \beta^{HWS}(u_{i+1}-\overline{u})(v_{j+1}-\overline{v})(w_k-\overline{w})]$$

$$- [\beta^{HW}(u_i-\overline{u})(v_{j+1}-\overline{v}) + \beta^{HWS}(u_i-\overline{u})(v_{j+1}-\overline{v})(w_k-\overline{w})$$

$$+ \beta^{HW}(u_{i+1}-\overline{u})(v_j-\overline{v}) + \beta^{HWS}(u_{i+1}-\overline{u})(v_j-\overline{v})(w_k-\overline{w})]$$

$$= [\beta^{HW} + \beta^{HWS}(w_k-\overline{w})] \times [(u_i-\overline{u})(v_j-\overline{v}) + (u_{i+1}-\overline{u})(v_{j+1}-\overline{v})$$

$$- (u_i-\overline{u})(v_{j+1}-\overline{v}) - (u_{i+1}-\overline{u})(v_j-\overline{v})]$$

$$= [\beta^{HW} + \beta^{HWS}(w_k-\overline{w})] \times [(u_{i+1}-u_i)(v_{j+1}-v_j)]$$

$$= \beta^{HW} + \beta^{HWS}(w_k-\overline{w}) \tag{2.13}$$

for unit-spaced scores $\{u_i\}$ and $\{v_j\}$. Odds ratio (2.13) can also be expressed as Agresti (1984) pointed out:

$$\log \Theta_{ij(k)} = \beta^{HW} + \beta^{HWS}[k - \tfrac{1}{2}(l+1)]$$

where k is the kth category and l is the number of layers for the third variable, sharing of housework (S). In (2.13), $(w_k - \overline{w})$ are equal to scores $(-1, 0, +1)$ that correspond to the three categories of housework sharing variable. In the above formulation, $[k - \tfrac{1}{2}(l+1)]$ yields the identical scores when $k = 1$, 2, or 3, and $l = 3$ are applied. This means that within a particular level of S, the association between H and W is uniform with local log odds ratio (2.13). Thus the strength of H-W partial association is constant within levels of S but changes linearly across the levels of S.

Other two-factor associations (H-S and W-S) in the uniform interaction model (2.10) are expressed and interpreted similarly:

$$\log \Theta_{i(j)k} = \beta^{HS} + \beta^{HWS}(v_j-\overline{v}),$$

and

$$\log \Theta_{(i)jk} = \beta^{WS} + \beta^{HWS}(u_i-\overline{u}).$$

If $\beta^{HWS} = 0$ there is no three-factor interaction and uniform interaction model (2.10) simplifies to homogeneous uniform association model (2.9).

Heterogeneous Uniform Association Model

In some cases, we may wish to regard one of the variables (say sharing of housework) as a control variable, and assume a uniform conditional association between husband's and wife's gender ideology that may change in an unspecified manner across the levels of sharing housework (S). In this approach, the interaction is not uniform, and the variable (S) is treated as nominal. This model can be expressed as:

$$\log F_{ijk} = \mu + \lambda_i^H + \lambda_j^W + \lambda_k^S + \lambda_{ik}^{HS} + \lambda_{jk}^{WS}$$

$$+ \beta_k^{HW} (u_i - \bar{u})(v_i - \bar{v}) \qquad (2.14)$$

where $\sum_i \lambda_i^H = \sum_j \lambda_j^W = \sum_k \lambda_k^S = \sum_k \beta_k^{HW} = 0$. This model implies that there is a uniform association between H and W within each level of S, but it also implies that the degree of that association is heterogeneous among levels of S. Stated another way, the relationship between husband's and wife's gender ideology is the same for respondents in each category but varies across categories of the housework variable. Using our example, this means that for couples who rarely share housework, the odds of both a husband and wife having a liberal gender ideology may be low. On the other hand, for those couples who frequently share housework, such odds may be higher. This shows more consistency of gender ideology between husbands and wives for those who share more housework than for those who rarely do.

The local log odds ratios, $\log \Theta_{ij(k)}$, of this model can be obtained as follows:

$$\log \frac{F_{ijk} F_{i+1,j+1,k}}{F_{i,j+1,k} F_{i+1,j,k}} = \beta_k^{HW}[(u_i - \bar{u})(v_j - \bar{v}) + (u_{i+1} - \bar{u})(v_{j+1} - \bar{v})$$

$$- (u_i - \bar{u})(v_{j+1} - \bar{v}) + (u_{i+1} - \bar{u})(v_j - \bar{v})]$$

$$= \beta_k^{HW}(u_{i+1} - u_i)(v_{j+1} - v_j)$$

$$= \beta_k^{HW}$$

Unlike Model 2.10, Model 2.14 does not assume uniform H-S and W-S partial associations. For comparable models with uniform H-S or W-S associations, local log odds ratios are, $\log \Theta_{i(j)k} = \beta_j^{HS}$ and $\log \Theta_{(i)jk} = \beta_i^{WS}$, respectively.

Multidimensional Log-Multiplicative Models

The RC model can also be generalized in various ways so that it can be used with multidimensional cross-classifications of ordinal variables. We can obtain log-multiplicative models when parameters are substituted for some or all pairs of sets of fixed scores in the ordinal log-linear models. For example, a parameter-score version of the partial association model (2.7) can be written:

$$\log F_{ijk} = \mu + \lambda_i^A + \lambda_j^B + \lambda_k^C + \beta^{AB}\mu_i\nu_j + \beta^{AC}\mu_i\omega_k + \beta^{BC}\nu_j\omega_k \quad (2.15)$$

where $\sum_i \lambda_i^A = \sum_j \lambda_j^B = \sum_k \lambda_k^C = 0$. The score parameters also satisfy constraints such as $\sum_i \mu_i = \sum_j \nu_j = \sum_k \omega_k = 0$. The model has df $= IJK - 2(I+J+K) + 5$ and is always unsaturated. Log odds ratios, $\Theta_{ij(k)}$, in Model 2.15 are

$$\log \frac{F_{ijk}F_{i+1,j+1,k}}{F_{i,j+1,k}F_{i+1,j,k}} = \beta^{AB}(\mu_i\nu_j + \mu_{i+1}\nu_{j+1} - \mu_i\nu_{j+1} - \mu_{i+1}\nu_j)$$

$$= \beta^{AB}(\mu_{i+1} - \mu_i)(\nu_{j+1} - \nu_j)$$

Other odds ratios are similarly derived:

$$\log \Theta_{i(j)k} = \beta^{AC}(\mu_{i+1} - \mu_i)(\omega_{k+1} - \omega_k),$$

$$\log \Theta_{(i)jk} = \beta^{BC}(\nu_{j+1} - \nu_j)(\omega_{k+1} - \omega_k),$$

and

$$\log \Theta_{ijk} = 0.$$

Thus the interpretation of this model is quite simple when the parameter scores are found to be monotonic.

This model estimates scores for all the ordinal variables. As in the two-way RC model, the odds ratio for two variables depends on the estimated scores. In Model 2.15, these odds ratios do not vary across the levels of the third variable. If a uniform association model fits poorly, but a model such as (2.15) fits well, the estimated parameter scores demonstrate non-monotonicity or are simply stronger over certain regions than over others.

Odds Ratios for Three-Way Log-Linear Models: Summary

Table 2.12 summarizes goodness of fit and local log odds ratios for ordinal log-linear models with three variables, husband's and wife's gender ideology, and sharing of housework. In the partial association model (2.7), we assume that a husband's ideology is measured at a nominal level. Then the odds of a wife holding a traditional (as opposed to moderate), or a moderate (as opposed to liberal) gender ideology vary depending on the level of a husband's ideology. The odds of rarely (as opposed to sometimes) or the odds of sometimes (as opposed to frequently) sharing housework also vary across levels of a husband's gender ideology. The W-S association between wife's ideology and housework, however, can be described by the uniform odds ratio, β^{WS}.

In the homogeneous uniform association model (2.9), local conditional log odds ratios simplify to β parameters. These odds ratios are uniform for each pair of variables, and the strength of association is homogeneous across the levels of the third variable. In the uniform interaction model (2.10), the local log odds ratios for two-factor association terms vary by level of the third variable because we need to add the $\beta^{HWS}(w_k - \overline{w})$, $\beta^{HWS}(v_j - \overline{v})$, and $\beta^{HWS}(u_i - \overline{u})$ factors for β^{HW}, β^{HS}, and β^{WS}, respectively. The local log odds ratio is constant and equals β^{XYZ} for all $2 \times 2 \times 2$ subtables formed for adjacent rows, adjacent columns, and adjacent layers.

The most parsimonious model in Table 2.12 is the homogeneous uniform association model, which yields a constant odds ratio for each second-factor association. The ordinal log-linear models also allow us to examine three-factor interaction terms, which is not possible with nominal models. This section has given a sampling of the types of models that can be fitted to multidimensional tables having ordinal classifications. For further discussions of these and other models, see Clogg (1982b), and Agresti and Kezouh (1983).

TABLE 2.12
Estimated Local Log Odds Ratios for Ordinal Log-Linear Models
Fitted to Table 2.10

Goodness of Fit	Partial Association (2.7)	Homogeneous Uniform Association (2.9)	Uniform Interaction (2.10)
G^2	19.83	20.28	20.23
df	15	17	16
Log Odds Ratio			
H-W association			
$\log \Theta_{ij(k)}$	$\tau_{1j}^{HW} = -0.639$ $\tau_{2j}^{HW} = 0.015$ $\tau_{3j}^{HW} = 0.624$	$\beta^{HW} = 0.895$	$\beta^{HW} + \beta^{HWS}(w_k - \overline{w}) =$ 0.893 for $k = 1$ 0.925 for $k = 2$ 0.957 for $k = 3$
H-S association			
$\log \Theta_{i(j)k}$	$\tau_{1k}^{HS} = -0.296$ $\tau_{2k}^{HS} = -0.038$ $\tau_{3k}^{HS} = 0.334$	$\beta^{HS} = 0.438$	$\beta^{HS} + \beta^{HWS}(v_j - \overline{v}) =$ 0.436 for $v = 1$ 0.468 for $v = 2$ 0.500 for $v = 3$
W-S association			
$\log \Theta_{(i)jk}$	$\beta^{WS} = 0.489$	$\beta^{WS} = 0.488$	$\beta^{WS} + \beta^{HWS}(u_i - \overline{u}) =$ 0.492 for $u = 1$ 0.524 for $u = 2$ 0.556 for $u = 3$
H-W-S association			
$\log \Theta_{ijk}$	$\beta^{HWS} = 0$	$\beta^{HWS} = 0$	$\beta^{HWS} = 0.032$

Selection for Ordinal Log-Linear Models

As demonstrated in this chapter, when we allow ordinal associations and analogous effects for interactions, the variety of potential models is much greater than the nominal models discussed in Chapter 1. It is thus necessary to develop a strategy for selecting among the models. Listed below is a summary of these strategies:

1. A parsimonious model containing the fewest parameters is preferable.
2. A model having a straightforward and reasonable interpretation is preferred. For example, "simply interpreted" in a four-way table may mean "no four-factor interaction term."
3. A model that contains all significant terms and no nonsignificant terms is preferable. The significant terms mean that the deletion of them from the model results in a model that fits significantly worse. The terms are

nonsignificant if the model formed by adding that term does not fit significantly better.

4. A high probability is desirable, but too high of a probability may suggest the model is overspecified. That is, though the model might fit the current sample, it is unlikely to fit another similar sample. This can arise when terms are included due to Type I errors occurring in the specification search.

5. For ordinal models, Agresti (1990) recommends that we start with a nominal log-linear model and find the model that fits well. Then we replace some parameters with structured terms for ordinal classifications.

6. Perhaps most important, let theory guide the analysis. Relying solely on G^2's generated by a computer program may lead to the acceptance of a final model that has little substantive importance. This criterion is crucial in the selection of both nominal and ordinal log-linear models. In selecting the model, we should not rely solely on statistical criteria.

Advantages of Using Ordinal Log-Linear Models

There are many advantages in using ordinal methods instead of the nominal log-linear models. First, as shown in our examples, where nominal models are saturated, there are unsaturated ordinal log-linear models. A simple example is the inclusion of two-factor association terms that make the nominal model saturated but not ordinal models. The ordinal models have structured association and interaction terms that contain fewer parameters, and therefore retain more residual degrees of freedom than the nominal models. Therefore, the ordinal technique provides a greater variety of models between independence and saturated models than nominal modeling. Second, the parameters in the ordinal models describe types of trends and are simpler to interpret than those in nominal models. Using odds ratios, we can easily describe the association among ordinal variables in the hypothesized model. Third, tests based on ordinal models have improved power for detecting certain types of association. Finally, the ordinal models allow a greater variety of models by including theoretically derived odds ratios for the hypothesized model.

Summary

The association parameters in the ordinal log-linear models describe ordinal characteristics of the data. Hence descriptive statements made with these methods are almost always more informative than those based on methods that ignore the ordinal nature of the variables. Generally, when the methodology recognizes inherent ordinality, there is greater power for detecting certain forms of association and there is a greater variety of ways of describing the association.

There is a large variety of models that can be used to analyze ordinal categorical data. This chapter focused on log-linear and log-multiplicative models for such data which provide more informative interpretations of ordinal variables than nominal log-linear models.

3. CONCLUSION

This monograph has presented models for analyzing cross-classified tables containing ordinal variables. These models are powerful statistical models. Their ability to detect the monotone trend of ordinal variables and to provide for more variety with respect to nonsaturated models gives them a major advantage over log-linear models for nominal variables. We have also discussed log-multiplicative models that use parameter estimates instead of assigning unit-spaced scores to ordinal variables. In this concluding chapter, I will emphasize three important issues for social scientists who try to apply the models described in this monograph.

First, the examples presented throughout the monograph are quite simple and the interpretation of parameters is straightforward. Although the most complex model presented in this monograph is based on a three-way table, the reader should not be left with the impression that more complex models cannot be analyzed. Complexity can be added by including more variables. It is, however, misleading to think the more complex the model, the better model it is. As previously discussed, inclusion of more variables quickly adds complexity, and problems with zero cells are most likely to occur. The reader needs to be constantly reminded of the importance of parsimony in log-linear model building.

Second, even though log-linear models have become widely accepted as a tool for analyzing complex relationships among categorical data, their full power has not always been exploited by researchers. Many applications of log-linear models make little interpretive use of parameter values despite the fact that log-linear methodology is a parametric technique should be one of its greatest strengths. The confusion and an attempt to correct errors regarding parameter interpretation has generated considerable attention in the log-linear literature (e.g., Long, 1984; Kaufman & Schervish, 1986). The full advantage of log-linear models cannot be realized until researchers begin to interpret the coefficients using odds ratios. The recent literature (e.g., Alba, 1988; Clogg & Eliason, 1988) on the interpretation of log-linear parameters is in the right direction toward full utility of this modeling technique.

Third, in a related vein, it is frequently the case that when log-linear models are used, researchers tend to pay too much attention to the fit of the model. This is partly due to the availability of rather simple and inexpensive computer specification search methods. Therefore, researchers spend too much time in their attempt to find the best fitting model. As described in this monograph, the best fitting does not necessarily imply the model that has the highest probability value. It is important to have a model that fits the data well. However, relying solely on G^2's and their corresponding probability values may result in the selection of a model that may not make much substantive sense. To this end, one cannot overemphasize the importance of theory and the literature that can provide a baseline model and shed light on some of the potentially important relationships to be included in the final model. If the model that fits the best does not contain the theoretically important relationship, the researcher may want to include such a term although the fit of the model may not be as good as the computer-driven model.

Methods for categorical data are constantly being upgraded and changed. Although a detailed presentation of these important advances is beyond the scope of this monograph, it is recommended that interested readers consult some of the primary sources for ordinal categorical data analysis listed below:

1. Goodman (e.g., 1979, 1981a, 1981b, 1981c, 1983) has written extensively on ordinal log-linear models and log-multiplicative models.

2. Agresti's books (1984, 1990) are extremely useful texts for ordinal categorical data. Though his new book (1990) is a more general text on categorical data analysis, the earlier book (1984) focuses on ordinal categorical data.

3. Clogg and others (1982a; Clogg & Eliason, 1988) have written extensively on ordinal log-linear models and other models that use parameter estimates. They present how these techniques can be applied to social science data.

56

4. Other recent applications of these models can be found in Clogg (1982b), Ishii-Kuntz (1991), Wong (1990), and Yamaguchi (1987). The reader is strongly encouraged to consult these sources. The best way to obtain a clear understanding of the ordinal methods described in this monograph is to study applications and to apply the model to actual data. These references are extremely useful in both respects.

Finally, all the models described in this monograph are a powerful set of statistical models. This, however, does not imply that other techniques being developed to study ordinal variables should be easily dismissed. Prior to the development of these model-based techniques, researchers relied heavily on measures of association. These measures include Gamma, Somer's *d*, Kendall's tau-*b* and tau (for details, see *Measures of Association* by Liebetrau, 1983, in the QASS series.) Unlike the models discussed in this monograph, these measures of association generally transform the cell proportions to a single number that describes a certain aspect of the association in the table. Thus a measure of association can be informative in giving us an indication of the strength of association in the table, but it does not give us as much information as a model. Perhaps the combination of these techniques may be extremely useful in many situations.

Most measures of association are quite simple to interpret and can be understood by a wide audience. Even if a simple model fits a table well, it usually enhances our understanding of the degree of association if we are given statistics such as the relative number of concordant and discordant pairs. Also, for some cross-classifications none of the commonly used models provide an adequate fit, even if levels of the explanatory variables are stochastically ordered (see Agresti, 1981). Measures of association that compare stochastically ordered distributions can be used whenever the log-linear row effects, logit row effects, or various other models are appropriate. If different model types fit different tables, measures of association give us a common basis for comparing associations and summarizing the results of the models.

APPENDIX:
COMPUTER SOFTWARE FOR ESTIMATING
ORDINAL LOG-LINEAR MODELS

This appendix summarizes statistical computer software available for ordinal log-linear and log-multiplicative models. Because there is a continual introduction of new programs and updating of existing ones, this summary is intended to be approximate. First, I will present several computer packages that can perform categorical data analysis. Second, I will present the sample programs that were used to fit the uniform association model described in Chapter 2.

Computer Software

SPSS. The LOGLINEAR program, described by Clogg and Becker (1986) and Norusis (1988), provides maximum likelihood (ML) fitting of a wide variety of log-linear models. The Advanced Statistics Guide describes log-linear models for ordinal data. In this program, the DESIGN statement specifies the form of the model. Optional output includes fitted values, standardized and adjusted residuals, and model parameter estimates and their correlations. It fits models having structural zeros by attaching zero weights to certain cells. The LOG-LINEAR is also available on their PC software, SPSS/PC+.

SAS. The main SAS procedure for ordinal log-linear models is CATMOD. This procedure, developed by Stanish (1986), is useful for building a wide variety of models for categorical data. The user can write a formula to construct a particular response for a model, or request one of the standard options. For further details on CATMOD, see Imrey (1985) and Stanish (1986).

GLIM. The interactive program GLIM (Numerical Algorithms Group, 1986) is a low-cost package for a main frame or PC, sponsored by the Royal Statistical Society. It is designed for ML fitting of generalized linear models. For further discussion of the use of GLIM in fitting generalized linear models, see Aitken, Anderson, Francis, and Hinde (1989), Healy (1988), and Lindsey (1989).

CDAS. CDAS is a new PC program written by Scott Eliason that can analyze a wide variety of categorical data, including RC models. By changing a program command, CDAS allows the use of major computer programs for categorical data such as ANOAS, ECTA, FREQ, and MLLSA.

ANOAS. ANOAS is a program written by Clifford Clogg for ML estimation of ordinal log-linear models and more general association models for two-way tables. It can fit the uniform association model, row and column effects models, the RC model, and many other models.

GAUSS. The log-linear analysis module, written by J. Scott Long, gives ML fitting using the Newton-Raphson method. The user can enter the model matrix, which permits fitting a wide variety of log-linear models, including models for ordinal variables or models that incorporate cell weights.

Sample Computer Programs

The following are sample programs used to estimate the uniform association model presented in the monograph. Lower case words can be replaced with other

58

variable names. These sample programs are only one way to use the available programs. Other options are also available and described in detail in each computer software instruction manual.

SPSS (LOGLINEAR), SAS (CATMOD, ML option), GLIM, and CDAS can fit all ordinal log-linear models presented in this chapter, using the Newton-Raphson method. The uniform association Model (2.3) is fitted to Table 2.3 using SPSS in Tables A.1 and A.2, using SAS in Table A.3, using GLIM in Table A.4, and using CDAS in Table A.5.

There are two ways to estimate the uniform association model using SPSS. In the first program (Table A.1), we specify ordinal variables by requesting orthogonal polynomial contrasts, as in Haberman (1974). The ordinal scores used by this program are the coefficients of orthogonal polynomials. For instance, for three categories of gender ideology in Table 2.3, it used $(-1/\sqrt{2}, 0, 1/\sqrt{2})$. Therefore, if parameter estimates are requested for these data, they will be a factor of $\sqrt{2}$ larger than those reported in Chapter 2, which were based on scores $(-1, 0, 1)$. Agresti (1984) also explains this program in more detail.

The second program, shown in Table A.2, is described in SPSS Advanced Statistics Guide (Norusis, 1988). In this program, the statement COMPUTE X= sets up the cross-product scores for the uniform association model, and the statement DESIGN= fits the model. The statement COMPUTE Y= sets up scores for the classification, and the statement DESIGN= fits the model. See Norusis (1988) for further details.

TABLE A.1
SPSS Used to Fit Uniform
Association Model: Polynomial Contrast

```
DATA LIST LIST / religion gender FREQ *
VALUE LABELS religion 1 'religious' 2 'moderate' 3
        'not religious'/ gender 1 'traditional' 2
        'moderate' 3 'liberal'
WEIGHT BY FREQ
LOGLINEAR religion gender (1,3)/
   PRINT = DEFAULT ESTIM/
   CONTRAST (religion) = POLYNOMIAL/
   CONTRAST (gender) = POLYNOMIAL/
   DESIGN = religion gender religion by gender(1)/
BEGIN DATA
1 1 466
: : :
3 3 160
END DATA
```

TABLE A.2

SPSS Used to Fit Uniform
Association Model: Cross-Product Scores

```
DATA LIST LIST / religion gender FREQ *
VALUE LABELS religion 1 'religious' 2 'moderate' 3
       'not religious'/ gender 1 'traditional' 2
       'moderate' 3 'liberal'
COMPUTE sreligin=religion
COMPUTE sgender=gender
COMPUTE X=sreligin*sgender
WEIGHT BY FREQ
LOGLINEAR religion gender (1,3) WITH sreligin
       sgender X/
  PRINT = DEFAULT ESTIM/
  DESIGN = religion gender X/
BEGIN DATA
1 1 466
: : :
3 3 160
END DATA
```

TABLE A.3

SAS Used to Fit Uniform Association Model

```
INPUT religion gender CARDS
1 1 466 1 2 317 1 3 203
: : :    : : :     : : :
3 1 51 3 2 120 3 3 160
PROC CATMOD ORDER=DATA; WEIGHT COUNT;
POPULATION religion;
MODEL gender = (1  0 -2,  0  1 -1,
        1  0 -4,  0  1 -2,
        1  0 -6,  0  1 -3)/ ML NOGLS PRED=FREQ;
```

In SAS (CATMOD), the default responses are the baseline-category logits. The log-linear models are fitted by specifying the model matrix for a corresponding generalized logit model. For instance, the model matrix in Table A.3 has size 6×3, because there are two logits in each of the three rows of Table 2.1 ($2 \times 3 = 6$), and because there are three parameters in the model for $\log (F_{ij}/F_{i3})$ implied by the uniform association model. The first two elements in each row of the model matrix pertain to the intercept parameters for the two logits.

60

TABLE A.4
GLIM Used to Fit Uniform Association Model

```
$UNITS 9
$FACTOR religion 3 gender 3
$DATA religion gender COUNT
$READ
1 1 466 1 2 317 1 3 203
: : :   : : :   : : :
3 1 51 3 2 120 3 3 160
$CALCULATE X = religion*happy $
$YVAR COUNT
$ERROR POIS
$FIT religion + gender + X $
```

TABLE A.5
CDAS Used to Fit Uniform Association Model

```
OUTFILE uniform.out
PROG ANOAS
TABLE 3 3
DATA
466 317 203
:    :    :
051 120 160
TITLE religiosity by gender ideology: uniform
      association
MODEL 2
EXECUTE
FINISH
BEEP
```

In GLIM, the $CALCULATE directives calculate the scores for use in the uniform association models. Alternatively, we could enter data vectors containing the scores, or use the $ASSIGN directive to create score vectors (see Table A.4).

In CDAS, Prog ANOAS can be used to fit various models for two-way tables. Uniform association models can be fit using the MODEL 1 subcommand. The odds ratios and the log odds ratios can be printed when specifying OPTIONS 3 (see Table 5.A). Other models can be fitted easily using MODEL statements.

For row effects (or column effects) models, these sample programs can be generalized easily. Instead of defining both variables as ordinal in the several ways listed above, only the ordinal variable needs such a definition.

REFERENCES

AGRESTI, A. (1980) "Generalized odds ratios for ordinal data." Biometrics 36: 59-67.
AGRESTI, A. (1981) "Measures of nominal-ordinal association." Journal of the American Statistical Association 76: 524-529.
AGRESTI, A. (1984) Analysis of Ordinal Categorical Data. New York: John Wiley.
AGRESTI, A. (1990) Categorical Data Analysis. New York: John Wiley.
AGRESTI, A., and KEZOUH, A. (1983) "Association models for multidimensional cross-classifications of ordinal variables." Communications in Statistics A12: 1261-1276.
AITKEN, M., ANDERSON, D., FRANCIS, B., and HINDE, J. (1989) Statistical Modeling in GLIM. Oxford: Clarendon Press.
ALBA, R. D. (1988) "Interpreting the parameters of log-linear models," in J. S. Long (ed.) Common Problems/Proper Solutions (pp. 258-287). Newbury Park, CA: Sage.
ANDERSEN, E. B. (1980) Discrete Statistical Models with Social Science Applications. Amsterdam: North-Holland.
BECKETT, J. O., and SMITH, A. D. (1981) "Work and family roles: Equalitarian marriage in black and white families." Social Service Review 55: 314-326.
BISHOP, Y. M. M., FIENBERG, S. E., and HOLLAND, P. W. (1975) Discrete Multivariate Analysis: Theory and Practice. Cambridge: MIT Press.
CLOGG, C. C. (1982a) "Some models for the analysis of association in multiway cross-classifications having ordered categories." Journal of the American Statistical Association 77: 803-815.
CLOGG, C. C. (1982b) "Using association models in sociological research: Some examples." American Journal of Sociology 88: 114-134.
CLOGG, C. C., and BECKER, M. (1986) "Log-linear modeling with SPSS^x," in D. M. Allen (ed.) Computer Science and Statistics: Proceedings 17th Symposium on the Interface (pp. 163-169). Amsterdam: North-Holland.
CLOGG, C. C., and ELIASON, S. R. (1988) "Some common problems in log-linear analysis," in J. S. Long (ed.) Common Problems/Proper Solutions (pp. 226-257). Newbury Park, CA: Sage.
COCHRAN, W. G. (1954) "Some methods strengthening the common χ^2 tests." Biometrics 10: 417-451.
DEMARIS, A. (1992) Logit Modeling: Practical Applications. Sage University Paper series on Quantitative Applications in the Social Sciences, series No. 07-086. Newbury Park, CA: Sage.
EVERS, M., and NAMBOODIRI, N. K. (1977) "A Monte Carlo assessment of the stability of log-linear estimates in small samples." Proceedings of the American Statistical Association, Social Statistics Section. Washington, DC: American Statistical Association.

62

FIENBERG, S. E. (1980) The Analysis of Cross-Classified Categorical Data (2nd ed.). Cambridge: MIT Press.

GOODMAN, L. A. (1970) "The multivariate analysis of qualitative data: Interactions among multiple classifications." Journal of the American Statistical Association 65: 226-256.

GOODMAN, L. A. (1979) "Simple models for the analysis of association in cross-classifications having ordered categories." Journal of the American Statistical Association 74: 537-552.

GOODMAN, L. A. (1981a) "Association models and canonical correlation in the analysis of cross-classifications having ordered categories." Journal of the American Statistical Association 76: 320-334.

GOODMAN, L. A. (1981b) "Association models and the bivariate normal distribution in the analysis of cross-classifications having ordered categories." Biometrika 68: 347-355.

GOODMAN, L. A. (1981c) "Three elementary views of loglinear models for the analysis of cross-classification having ordered categories." Pp. 193-239 in Sociological Methodology. San Francisco: Jossey-Bass.

GOODMAN, L. A. (1983) "The analysis of dependence in cross classifications having ordered categories, using log-linear models for frequencies and log-linear models for odds." Biometrics 39: 149-160.

GRAUBARD, B. I., and KORN, E. L. (1987) "Choice of column scores for testing independence in ordered 2 × K contingency tables." Biometrics 43: 471-476.

GRIZZLE, J. E., STARMER, C. F., and KOCH, G. C. (1969) "Analysis of categorical data by linear models." Biometrics 26: 489-504.

HABERMAN, S. J. (1974) "Loglinear models for frequency tables with ordered classifications." Biometrics 30: 589-600.

HABERMAN, S. J. (1981) "Tests for independence in two-way contingency tables based on canonical correlation and on linear-by-linear interaction." Annals of Statistics 9: 1178-1186.

HANUSHECK, E. A., and JACKSON, J. E. (1977) Statistical Methods for Social Scientists. New York: Academic Press.

HAGENAARS, J. A. (1990) Categorical Longitudinal Data. Newbury Park, CA: Sage.

HEALY, M. J. R. (1988) GLIM: An Introduction. Oxford: Clarendon Press.

IMREY, P. B. (1985) "SAS software for log-linear models." In Proceedings of the 10th Annual SAS Users Group International Conference. Gary, NC: SAS Institute, Inc.

ISHII-KUNTZ, M. (1991) "Association models in family research." Journal of Marriage and the Family 53: 337-348.

ISHII-KUNTZ, M., and COLTRANE, S. (1992) "Predicting the sharing of household labor: Are parenting and housework distinct?" Sociological Perspectives 35: 629-647.

KAMO, Y. (1988) "Determinants of household labor: Resources, power, and ideology." Journal of Family Issues 9: 177-200.

KAUFMAN, R. L., and SCHERVISH, P. G. (1986) "Using adjusted crosstabulations to interpret log-linear relationships." American Sociological Review 51: 717-733.

KEITH, P., HILL, K., GOUDY, W. J., and POWERS, E. A. (1984) "Confidants and well-being: A note on male friendships in old age." The Gerontologist 24: 318-320.

KNOKE, D., and BURKE, P. J. (1980) Log-Linear Models. Sage University Paper series on Quantitative Applications in the Social Sciences, series No. 07-020. Beverly Hills and London: Sage.

63

KOEHLER, K., and LARNTZ, K. (1980) "An empirical investigation of goodness-of-fit statistics for sparse multinomials." Journal of the American Statistical Association 75: 336-344.

LANDALE, N. S., and GUEST, A. M. (1990) "Generation, ethnicity, and occupational opportunity in late 19th century America." American Sociological Review 55: 280-296.

LARNTZ, K. (1978) "Small sample comparison of exact levels for chi-squared goodness-of-fit statistics." Journal of the American Statistical Association 73: 253-263.

LIEBETRAU, A. M. (1983) Measures of Association. Sage University Paper series on Quantitative Application in the Social Sciences, series No. 07-032. Beverly Hills, CA: Sage.

LINDSEY, J. K. (1989) The Analysis of Categorical Data Using GLIM. New York: Springer.

LONG, J. S. (1984) "Estimable functions in log-linear models." Sociological Methods & Research 12: 399-432.

Numerical Alogorithms Group. (1986) The GLIM System Release 3.77 Manual. Downers Grove, IL: Numerical Alogorithms Group.

NORUSIS, M. J. (1988) SPSS^x Advanced Statistics Guide (2nd ed.) New York: McGraw-Hill.

O'BRIEN, R. M. (1979) "The use of Pearson's R with ordinal data." American Sociological Review 44: 851-857.

PINDYCK, R. S., and RUBINFELD, D. L. (1981) Econometric Models and Economic Forecasts. New York: McGraw-Hill.

STANISH, W. M. (1986) "Categorical data analysis strategies using SAS software," in D. M. Allen (ed.) Computer Science and Statistics: The Interface (pp. 239-256). New York: Elsevier North-Holland.

STOCKARD, J., and JOHNSON, M. M. (1980) Sex Roles: Sex Inequality and Sex Role Development. Englewood Cliffs, NJ: Prentice Hall.

SWEET, J., BUMPASS, L., and CALL, V. (1988) "The design and content of the National Survey of Families and Households." Working Paper NSFH-1, Center for Demography and Ecology, University of Wisconsin-Madison.

WILLIAMS, N. (1990) The Mexican American Family: Tradition and Change. Dix Hills, NY: General Hall.

WONG, R. S. (1990) "Understanding cross-national variation in occupational mobility." American Sociological Review 55: 560-573.

YAMAGUCHI, K. (1987) "Models for comparing mobility tables: Toward parsimony and substance." American Sociological Review 52: 482-494.

ZAVELLA, P. (1987) Women's Work and Chicano Families: Cannery Workers of the Santa Clara Valley. Ithaca, NY: Cornell University Press.

ABOUT THE AUTHOR

MASAKO ISHII-KUNTZ, Associate Professor at the University of California, Riverside, received her Ph.D. from Washington State University. She has published articles on the social psychological aspects of the family and on quantitative methods in such journals as the *Journal of Marriage and the Family, Journal of Family Issues,* and *Sociological Perspectives,* among others. She is currently working on a book that examines the diversity of Asian American families.